A GOODLY GALLERYE

A GOOD= LY GALLERYE

WITH A MOST PLEA= saunt Prospect, into the garden of naturall contemplation, to behold the naturall cau= ses of all kynde of *Meteors*, as wel fyery and ayery, as watry and earthly, of whiche sort be blasing sterres, shooting starres, flames in the ayre &c. thōder, lightning, earthquakes, &c. rayne dewe, snowe, cloudes, springes &c. stones, me= talles, earthes &c. to the glory of God, and the profit of his crea= turs.

¶ *PSALM.* 148.

Prayse the Lorde vpon earth Dragons and all deepes, Fyre, Haile, Snowe, Ise, Wyndes, and stormes, that doe his wyll.

LONDINI.
Anno. 1563.
(.·.)

MEMOIRS OF THE
AMERICAN PHILOSOPHICAL SOCIETY
Held at Philadelphia
For Promoting Useful Knowledge
Volume 130

A Goodly Gallerye

WILLIAM FULKE'S BOOK OF METEORS

(1563)

Edited with an Introduction
and Notes by

THEODORE HORNBERGER

THE AMERICAN PHILOSOPHICAL SOCIETY
INDEPENDENCE SQUARE • PHILADELPHIA
1979

Dr. Theodore Hornberger (1906–1975), Professor Emeritus at the University of Pennsylvania, was author and editor of many books on early American literature and the history of early Anglo-American science. He died before he could finish his edition of William Fulke's *A goodly gallerye,* which his friend and colleague Professor M. A. Shaaber has readied for publication.

Library of Congress Catalog Card Number 78-068390
International Standard Book Number 0-87169-130-2
US ISSN 0065-9738

Introduction

1. *The Author*

William Fulke (1538–1589) has a modest place in the history of Tudor Puritanism. Perhaps he was less radical in his maturity than in his youth, but for most of his life he was a leader in the bitter struggle to eradicate Romanist influences in the Church of which Elizabeth I was determined to be head.[1] For his services, which were as much political as doctrinal, Fulke was rewarded with the mastership of Pembroke Hall, Cambridge, where he reigned from 1578 until his death.[2]

As a young man, however, Fulke appears to have been less interested in religion and in law, to which he turned reluctantly to please his father, than in natural philosophy and mental games. These two relatively secular enthusiasms he managed to combine through the invention of two games played on squares, one based on astronomy and the other on geometry. *Ουρονομαχια, hoc est, astrologorum ludus* (London, 1571) describes pieces representing the sun, moon, and planets, which were to be moved about according to formulae relating to their various conjunctions and oppositions. These instructions, which were apparently printed at the time the game was offered for sale, were dedicated to William Cecil, Lord Burghley, then chancellor of Cambridge. *Μετρομαχια siue ludus geometricus* (London, 1578), dedicated to Robert Dudley, Earl of Leicester and, as will appear in a moment, Fulke's patron, at this date chancellor of Oxford, describes the moving of various elaborate geometrical figures toward a castle-like goal, in a way like some present-day children's games played with dice. Neither supplanted chess, but they both reveal a lively, inventive mind, oriented to the scientific interests of the age.

---

[1] M. M. Knappen, *Tudor Puritanism: a Chapter in the History of Idealism* (Chicago, 1939), p. 290 and *passim*.

[2] The chief authority on Fulke's life is Charles Henry and Thompson Cooper, *Athenae Cantabrigienses* **2** (Cambridge, 1861): pp. 57-61, 544, with an extensive bibliography. See also the article by the Rev. Canon Venables in the *Dictionary of National Biography* **20** (1889): pp. 305-308.

Born in London, Fulke received his early education in one of the city schools, either that supported by the Mercers' Company or, and more probably, St. Paul's. In 1555, during the troubled reign of Mary, he matriculated at St. John's College, Cambridge, where he received his B.A. degree in 1558 and his M.A. in 1563, the year in which *A Goodly Gallerye* was first printed. Between 1558 and 1564 he was reading law at Clifford's Inn. Whether he is the author of the "almanacke and pronostication of fulkes" entered in the Stationers' Register in 1560 by Henry Sutton cannot be ascertained since no copy has survived.[3] That same year saw in print an attack upon astrology which is still well known.[4] It appeared first in Latin, and soon after, in an English translation by William Painter, as *Antiprognosticon that is to saye, an Inuectiue agaynst the vayne and vnprofitable predictions of the Astrologians as Nostradame, &c. Translated out of Latine into Englishe. Wherunto is added by the author a shorte Treatise in Englyshe, as well for the vtter subuersion of that fained arte, as also for the better vnderstandynge of the common people, vnto whom the fyrst labour seemeth not sufficient* (*Short Title Catalogue* 11420). This title looks forward to two important aspects of *A Goodly Gallerye:* its rationalism and its central intention of making things clear to "the common people," that is to say, the unlearned populace now beginning to learn to read in their native language. This little book attacked astrology as a false science and asserted that astronomy was a true one.[5] Although Fulke noted in passing the impiety of presuming to foretell the acts of God, he relied more upon argument and previous authorities than upon a religious appeal.

Fulke's next publication was a book on another chess-like game based on arithmetical, geometrical, and musical proportions, here attributed to Pythagoras. The extent of his authorship remains a mystery. Two issues appeared in 1563, the second bearing the title of *The most noble auncient, and*

---

[3] Arber's *Transcript* **1**: p. 153.

[4] See Don Cameron Allen, *The Star-crossed Renaissance: the Quarrel about Astrology and its Influence in England* (Durham, N.C., 1941), pp. 106-112.

[5] There is no need for Lynn Thorndike's sneer: "since the names mentioned are mainly those of popular and sensational rather than learned and would be scientific astrologers, perhaps his [Fulke's] objection was more to such practitioners than to the art" (*A History of Magic and Experimental Science* **6** [New York, 1941]: p. 180).

learned playe, called the Philosophers game, inuented for the honest recreation of students, and other sober persons, in passing the tediousnes of tyme, to the release of their labours, and the exercise of their wittes. Set forth with such playne precepts, rules, and tables, that all men with ease may vnderstande it, and most men with pleasure practise it, by Rafe Leuer and augmented by W. F. (S.T.C. 155422). This was dedicated to Leicester, in verse, by the bookseller, James Rowbotham, has sometimes been attributed to William Fulwood, and was apparently an embarrassment to Fulke and to Ralph Lever, both of whom later claimed it.[6] The dedication to *A Goodly Gallerye* gives Fulke's side of the story. The latter must have been written not long after the appearance of the book printed for Rowbotham.

In 1564 Fulke was made a fellow of his Cambridge college, St. John's, and for the remainder of his life he was bound up with the interests of the university. These were religious rather than scientific, and there is little in his voluminous later writings —some thirty titles altogether—to remind us of his youthful enthusiasms. He seems to have given up the law thankfully, along with his inheritance, and to have subjected himself willingly to the all-absorbing concerns of university and church.

Among his associates at Cambridge was Thomas Cartwright, one of the most influential of Puritan leaders. Fulke did not follow Cartwright into Presbyterianism, but in 1565 he played a leading role in the "vestiarian controversy," dramatically refusing to wear the surplice prescribed by university custom. For this rebellion he was expelled from his fellowship. His friends rallied around him, and within a year he was reinstated. He took the B.D. degree in 1568 and soon thereafter was suspected of connivance in a marriage which by the laws of the church was technically incestuous. He resigned his fellowship under pressure, but was acquitted of the charge against him and again readmitted. In 1569, at the age of thirty-one, he was seriously in the running for the vacant mastership of his college, having, it is said, the active support of Leicester. Failing of election, he was appointed Leicester's private chaplain, and

---

[6] For an account of the confusion about this book see Eleanor Rosenberg, *Leicester, Patron of Letters* (New York, 1955), pp. 39-42.

the earl also secured for him the livings of Warley in Essex and Dennington in Sussex, which he retained for the remainder of his life. In 1572, by royal mandate, and again through Leicester's influence, he was given a D.D. degree, and in 1578 was chosen master of Pembroke Hall. There he lived out his life, not without incurring some accusations of greed and ambition, but serving the church as preacher and polemicist, proving Rome to be Babylon and confuting a whole series of Romanist controversialists with a vigor and violence remarkable even in an age when such qualities were the rule of debate rather than the exception.

2. *A Goodly Gallerye: Publishing History*

A Goodly Gallerye was entered in the Stationers' Register by William Griffith during the Stationers' fiscal year 1562–1563.[7] Griffith was a publisher in a small way whose list as a whole suggests that he published for the common sort of reader, for whose needs Fulke avowedly catered in his book. Griffith's edition of the work, an octavo of 76 leaves, is dated 1563 (*S.T.C.* 11435). The author's name ("William Fulce") appears in the dedication and his initials at the end. Eight years later Griffith reissued the unsold sheets with a new title page dated 1571 and a new setting of the dedicatory epistle (*S.T.C.* 11436). This reissue is the last book Griffith is known to have published; very possibly he died not long afterwards and the *Goodly Gallerye,* for which the demand does not seem to have been brisk anyway, became an orphan.[8] But on 23 July 1601, it was entered in the Stationers' Register by Simon Stafford, an enterprising printer not averse to picking up a derelict of potential value, as "an old copie printed by William Griffyth."[9] Stafford printed it the next year but transferred the copyright to William Leake, who entered "Doctor fulkes meteors" on 2 August 1602,

[7] Arber **1**: p. 213.

[8] A 1601 edition is reported in the *S.T.C.* (11437), locating a single copy at Corpus Christi College, Oxford. I am told by Mr. P. J. Law, writing for the librarian, that this is an error; the Corpus Christi copy is dated 1602 and belongs with *S.T.C.* 11438.

[9] Arber **3**: p. 189.

"by consent of Simon Stafford."[10] In this edition (*S.T.C.* 11438), the first of four published by Leake and his son, namesake, and successor, the title was slightly abbreviated, the author's name was put on the title page, the dedication was dropped, and a table of contents was added, but no substantive changes were made in the text. In 1634 the remaining sheets of the 1602 edition were reissued with a cancel title page which describes it as "The second Edition corrected and amended" (*S.T.C.* 11439). The publisher's name does not appear on the cancel title page. Leake had retired in 1618 and died in 1633. His son and namesake had been made free of the Stationers' Company in 1623 and was an active publisher after 1636, but what business he carried on before 1634 is unknown. But certainly no one but Leake's widow or his son is likely to have been in possession of unsold sheets of the 1602 printing and the subsequent transfer of the copyright from the widow to the son shows that ownership remained in the family. The imprint, "Printed by *Iohn Haviland,* and are to be sold by *Iames Boler,"* may be interpreted to mean that the distribution of the book was deputed to Boler by either the widow or the son, an arrangement which would admit of the omission of the publisher's name. Actually Haviland could have printed no more than the new title page. After her husband's death the Widow Leake, on 1 June 1635, assigned a number of copies, among which "Doctor ffulkes Meteors" is specifically mentioned, to his son.[11] The latter launched another edition, described as the third, in 1639 (*S.T.C.* 11440) ; most surviving copies have a title page dated 1640 (*S.T.C.* 11441). The next edition, dated 1654 (Wing F2260), alters the title to *Meteors: or, a plain description of all kinds of meteors,* etc., and reduces the author's name to initials. Copies dated 1655 (Wing F2260A) are also found. To this edition a "person of quality" identified only as "F. W." added fifteen pages of "Observations on Dr. F. his Booke of Meteors" which collect some information about metals and minerals. The last edition, dated 1670 (Wing F2261), prints the same text as all the others.

---

[10] Arber **3:** p. 213. It was not unusual for a printer to acquire a piece of copy and turn it over to a bookseller to market, the printing of it being presumably part, perhaps all, of the consideration for the transfer.

[11] Arber **4:** p. 340.

Thus Fulke's book was current for more than a hundred years, appearing in five editions with nine different dates without substantial change of text. The most likely reason for its durability is given by the publisher of the last edition thus: "I may (without breach of Modesty) affirm, that there is not in our Language any Booke of so small a Bulke, contains so much of the Doctrine of the *Meteors.*" The book's slow start and the concentration of six of the nine issues in the middle third of the seventeenth century is perhaps a measure of the increased interest in natural phenomena at that time.

3. *Description of Editions*

1563

Title: A | GOOD= | LY GALLERYE | *WITH A MOST PLEA* = | *saunt Prospect, into the garden of* | **naturall contemplation, to** | **be-hold the naturall cau=** | **ses of all kynde of** | *Meteors,* **as wel** | **fyery and** | **ayery, as watry and earthly, of whiche** | **sort be blasing sterres, shooting starres,** | **flames in the ayre &c. thōder, lightning,** | **earthquakes, &c. rayne dewe, snowe,** | **cloudes, springes &c. stones, me=** | **talles, earthes &c. to the** | **glory of God, and the** | **profit of his crea=** | **turs.** | ¶*PSALM.* 148. | *Prayse the Lorde vpon earth Dragons and all* | *deepes, Fyre, Haile, Snowe, Ise, Wyndes,* | *and stormes, that doe* | *his wyll.* | *LONDINI.* | *Anno. 1563.* | (∴)

Colophon: ❧ **Imprynted** | **at London in Fletestreate, at the** | **signe of the Faucone, by William** | **Griffith: And they are to be sold** | **at his shop in S. Dunstones** | **churchyarde in the** | **Weste.** | 1563.

Collation: 8°. π^4 (π1 blank) A-I^8 (I8 blank). Foliated 1-71 (A1-I7; 56 misnumbered 54).

Running heads: $\pi 3^v$-$\pi 4^v$ The Epistle. A1^r *A GOODLY GALLERY.* A1^v-17^r *A GOODLY* | *GALLERY.*

Anomalous catchwords: B7^r c.w. *lation* B7^v *tion* F4^r **ture, and** F4^v **and** G4^r **not long** G4^v **long** H6^r **truth of these** H6^v **thinges** I2^r **shion of** I2^v **of** I6^r **briefly & yet** I6^v **sufficiently** $\pi 4^v$, A2^v, A4^v, A6^r, B1^v, B8^r, C1^v, C8^r, E8^r catchwords wanting.

Type: Black letter, with italic for some headings, marginal notes, proper names, and Latin words.

Contents: $\pi 2^r$ title page. $\pi 3^r$ dedication. A1^r The first Booke. A6^v The seconde Booke of fyery *Meteores.* C1^v, l. 18 The thirde Booke of aery impressions. F6^v, l. 7 The fourth booke *of watry impressions.* H6^r The fift booke of earthly *Meteores* or bodies perfectly mixed. I7^v colophon.

Copies: British Museum, Huntington Library, University of Michigan (Dictionary Office). *S.T.C.* 11435.

1571

Title: [within a border of type ornaments] A goodly Gallery vvith | **a most pleasaunt Prospect, into** | the garden of naturall contem- | **plation, to beholde the na=** | *turall causes of all kind* | of Meteors. | **As well fyery and ayery, as watry** | **annd earthly, of which sorte be blasing** | **Starres, shootinge Starres, flames** | **in the ayre &c. thonder, Lightninge,** | **Earthquakes. &c. Rayne Dew, snowe** | **Cloudes, Springes. &c. Stones,** | **Metalles, Earthes. To the** | **glory of God, and the** | **profitte of his** | **creatures.** | ¶PSALME 148. | ¶Prayse the Lord vpon Earth, | Dragons & all depes, Fyre, Haile | Snowe, Ise, VVinds, and stormes | that do his will. | **Imprinted at London, in Fletestrete** | **by Wylliam Gryffith.** 1571.

Collation: 8°. π^4 (π 1 blank) A-I^8 (I8 blank). Foliated 1-71 (A1-I7).

Copies: Bodleian Library, British Museum, Huntington Library. *S.T.C.* 11436.

A reissue of sheets A-I of the 1563 printing with a new title page and a new setting of the dedication.

1602

Title: [within a border (McKerrow & Ferguson 86)] A most pleasant Prospect | into the Garden of naturall Con- | templation, to behold the na- | turall causes of all kind of | *Meteors:* | As well fiery and ayry, as watry and earthly: of | which sort be blazing Starres, shooting Starres, | flames in the ayre &c. Thunder, Lightning, | Earthquakes, &c. Rayne, Dewe, Snow, | Clouds, Springs, &c. Stones, Metals, | and Earths: To the glory of | God, and the profit of his | creatures. | By W. Fulke Doctor of Diuinitie. | *Prayse the Lord vpon earth, Dragons* | *and all deepes, Fyre, Hayle, Snowe,* | *Ice, Windes and stormes that doe* | *his will, Psal.* 148. | AT LONDON | Printed for William Leake, dwel-|ling in Paules Church-yard, at the signe of | the holy Ghost. 1602.

Collation: 8°. ¶[4] (¶1 blank) A-I[8] (I7[v], I8 blank). Foliated 2-71 (A2-I7).

Copies: British Museum, Cambridge University, Columbia University (Law Library), Corpus Christi College, Oxford, Folger Library, Harvard University, Huntington Library. S.T.C. 11438.

Except for occasional deviations, a page-for-page reprint. That the printer was Simon Stafford may be inferred from the title page border.

1634

Title: A | Most pleasant Prospect | Into the Garden of | Naturall Contemplation, to | behold the naturall causes of | all kinde of Meteors: | As well fierie and airie, as watrie and | earthly: of which sort the blazing Starres, | shooting Starres, Flames in the Aire, &c. Thunder, | Lightning, Earthquakes, &c. Raine, Dew, Snow, | Clouds, Springs, &c. Stones, Metals, and | Earths: To the glorie of God, and | the profit of his creatures. | By *W. Fulke* Doctor of Divinitie. | *Praise the Lord vpon earth, Dragons, and all Deepes,* | *Fire, Haile, Snow, Ice, Winds and Stormes that* | *doe his will.* Psal. 148. | The second Edition corrected and amended. | [rule] | *LONDON,* | Printed by *Iohn Haviland,* and are to be sold by | *Iames Boler* dwelling at the Marygold | in *Pauls* Church-yard. 1634.

Collation: 8°. ¶[4] (¶1 blank) A-I[8] (I7[v], I8 blank). Foliated 2-71 (A2-I7).

Copies: British Museum, Cambridge University, Edinburgh University, Folger Library, Glasgow University, Huntington Library, National Library of Medicine, University of Pennsylvania, University of Wisconsin. *S.T.C.* 11439.

Sheets of the 1602 printing with a cancel title page.

1639

Title: A | *Most pleasant Prospect.* | INTO THE GARDEN | of Naturall Contemplation, | to behold the naturall causes of | all kinde of Meteors. | As well fiery and airie, as watrie and |

earthly: of which sort the blazing Starres, | shooting Starres, Flames in the Aire, &c. | Thunder, Lightning, Earthquakes, &c. Raine, Dew, | Snow, Clouds, Springs, &c. Stones, Metals, and | Earths: To the glory of God, and the | profit of his creatures. | By *W. Fulke* Doctor of Divinity. | *Praise the Lord upon earth, Dragons, and all Deepes;* | *Fire, Haile, Snow, Ice, Windes and Stormes that doe* | *his will.* Psal. 148. | The third Edition corrected and amended. | [rule] | *LONDON.* | Printed by *E. G.* for *William Leake,* and are to be | sold at his shop betwixt the *Rowles,* and | *Sargents Inne.* 1639.

Collation: 8°. ❡[4] (❡1 blank) A-I[8] (I7[v], I8 blank). Foliated 2-71 (A2-I7).

Copies: British Museum (title page defective), Cambridge University, Folger Library. *S.T.C.* 11440.

Except for a few slight deviations, a page-for-page reprint of the previous edition.

1640

Title: identical to the preceding except for the imprint: *LONDON.* | Printed by *E. G.* for *William Leake,* and are to be | sold at his shop in *Chancery-lane* neere | the *Rowles.* 1640.

Collation: the same.

Copies: British Museum, Cambridge University, Detroit Public Library, Duke University, Folger Library, Harvard University, University of Illinois, John Rylands Library, University of North Carolina, University of Oklahoma, Peterborough Cathedral, Rutgers University, University of Texas, Wigan Public Library, Yale University. *S.T.C.* 11441.

Sheets of the 1639 printing with a variant title page.

1654

Title: [within a border of rules] Meteors: | OR, | A plain Descripti- | on of all kind of *Meteors,* | as well *Fiery* and *Ayrie,* as | *Watry* and *Earthy:* | BRIEFLY | Manifesting the *Causes* of | all *Blazing-Stars, Shooting-Stars,* | *Flames in the Aire, Thunder,* | *Lightning, Earthquakes,* | *Rain, Dew, Snow, Clouds,* | *Springs, Stones,* and | *Metalls,* | *By* W. F. *Doctor in* | *Divinitie.* | *LONDON,* | Printed for *William Leake* at the Crown | in *Fleet-street,* between the two Temple | Gates, 1654.

Collation: 8°. a[4] A-L[8]. Paginated 1-174 (A1[r]-L7[v]).

Copies: Bibliothèque Nationale, Paris, W. A. Clark Library (University of California at Los Angeles), Harvard University, University of Wisconsin. Wing F2260.

1655

Title: identical to the preceding except for the date: 1655.

Collation: the same.

Copies: Bodleian Library, Boston Public Library, British Museum, Congregational Library, London, Harvard University, Yale University. Wing F2260A.

1670

Title: [within a border of rules] Meteors; | OR, | A plain Description | of all kind of *Meteors,* | as well *Fiery* and *Ayrie,* as | *Watry* and *Earthy:* | BRIEFLY | Manifesting the *Causes* of | all *Blazing-Stars, Shooting-Stars,* | *Flames in the Aire, Thunder,* | *Lightning, Earthquakes,* | *Rain, Dew, Snow, Clouds,* | *Springs, Stones,* and | *Metalls.* | [rule] | By *W. F.* Doctor in | Divinity. | [rule] | *LONDON,* | Printed for *William Leake,* at the Crown | in *Fleet-street,* between the two | Temple Gates, 1670.

Collation: 8°. a^4 $A\text{-}L^8$. Paginated 1-174 ($A1^r$-$L7^v$).

Copies: Boston Public Library, British Museum, Cambridge University, W. A. Clark Library (University of California at Los Angeles), Library of Congress, Harvard University, University of Michigan, Princeton University, Yale University. Wing F2261.

A page-for-page reprint. Listed in *Mercurius litterarius for Easter term 1670* and in *A catalogue of books printed and published at London in Easter term, 1670 (Term catalogues,* ed. E. Arber, **1**: pp. 32, 39).

4. *Fulke's Main Sources*

Like all works of its kind, *A Goodly Gallerye* draws upon many sources and reflects speculation about the nature of the universe over hundreds and hundreds of years. Since it was the first book of its kind in English, its chief sources are Greek, Latin, and Arabic treatises, some of which Fulke probably knew at first hand. Very probably, however, he relied heavily upon the commentaries which the schoolmen of the Middle Ages had provided in large numbers, so that a precise description of his immediate indebtedness would require a fuller examination of obscure books than I am prepared to make. In broad outline, however, what he owed to the past is clear enough. He himself refers more than once to the triumvirate to whom he is most obligated: Aristotle, Seneca, and Pliny. In addition to these three, he cites in the course of his treatise twenty-five other authors, many of whom he probably knew only at second hand. The only contemporary whose work was of much importance to him was Jerome Cardan (1501-1575), the colorful author of *De Subtilitate Rerum* (1551), a book which Fulke evidently had read.

I shall leave the details of Fulke's indebtedness to his predecessors for a series of brief notes on the separate chapters of *A Goodly Gallerye*. To set the stage, however, some account of what the book owes to Aristotle, Seneca, and Pliny should be helpful.

Aristotle

The basic ideas of Fulke's book come from the scientific treatises attributed to Aristotle, his associates, or his pupils, among whom the schoolmen made little distinction. The *Physica* is relatively unimportant, but the greater part of *A Goodly Gallerye* was ultimately derived from the third and fourth books of *De caelo*, the second book of *De generatione et corruptione*, the *De mundo*, and the *Meteorologica*. The two last-named treatises were probably not written by Aristotle, but for some centuries they were accepted as part of the canon and studied assiduously in the schools.[12] *De caelo* and *De generatione et corruptione* contain the fundamental Aristotelian theory of the heavens, composed of an incorruptible celestial element, and the earth, made up of four corruptible elements, fire, air, water, and earth, each constantly changing and interacting with the others. *De mundo* describes the various spheres in which the four elements appear in their purest forms, together with the supposed effects of vapors and exhalations. The *Meteorologica*,

[12] The *De mundo* is ordinarily attributed to Posidonius.

in the first three of its four books,[13] applies the foregoing theories to great numbers of phenomena: shooting stars, comets, the Milky Way, rain, dew, frost, snow, hail, winds, rivers, springs, the tides and the saltness of the sea, earthquakes, the rainbow, and a dozen more. It is Fulke's chief source. His organization, however, is much more systematic, perhaps in accordance with previous scholastic practice, although I have not discovered a particular source for it.

By 1563 the Aristotelian corpus was readily available thanks to the first century of printing. The *Opera omnia* had been printed in Latin at least eleven times, the *Meteorologica* (by itself or with an accompanying commentary) at least twenty-eight. The commentary Fulke is most likely to have known is that by St. Thomas Aquinas, but he doubtless knew many others.

Seneca

Fulke refers eleven times to the younger Seneca (4 B.C.?-65 A.D.). These references, however, do not fully reveal the extent of his debt to the *Quaestiones naturales,* a book which deals with approximately the same subjects as the *Meteorologica.* What Seneca gave him was a wealth of literary and other allusions extremely useful to a writer seeking to appeal to the "common sort" of reader. From Seneca, moreover, Fulke seems to have derived his tone, which is one of questioning and weighing variant opinions, often leaving the reader to draw his own conclusions. The *Quæstiones naturales,* moreover, has a religious flavor which, although Stoic rather than Christian, is remarkably similar to that of *A Goodly Gallerye.*

During the Middle Ages, writes Seneca's best-informed translator,

> in default of any general knowledge of Aristotle, Seneca was the chief authority on Physical Science. The views transmitted by him, for they were comparatively seldom his own, having obtained currency, found their way into literature, and probably went far to

[13] Book IV, which deals with what would now be called chemical change, is only distantly related to *A Goodly Gallerye.*

colour the conceptions entertained on the subject in all the earlier literature of Modern Europe.[14]

The *Quaestiones naturales,* however, was seldom printed separately, and it drew to itself no such extensive body of commentary as that for Aristotle's writings. Between 1475 and 1563, however, Seneca's *Opera omnia* were printed at least ten times.

Pliny

Compared to the *Meteorologica* and the *Quaestiones naturales,* the *Historia Naturalis* of the elder Pliny (28-79 A.D) is an uncritical and unquestioning book.[15] Nevertheless, Fulke used it extensively to enliven *A Goodly Gallerye.* He makes only six direct references to Pliny, but his debt to him was as great, probably, as that to Seneca, so far as illustrative material goes. The most important portions of the *Historia Naturalis* to his purposes were Book II, which ranges from comets to earthquakes, and Books XXXIII-XXXIV, which treat of metals.

By the time Fulke wrote, at least thirty-seven Latin editions of the *Historia Naturalis* had been printed. Philemon Holland's famous translation into English was not to appear until 1601.

A Goodly Gallerye has been described as "an advanced treatise on the wonders of meteorology,"[16] but it is "advanced" chiefly in that it seeks to provide the new reading public of the English Reformation with knowledge which has previously been available only in Latin. It is essentially an elementary summary of ancient and medieval natural philosophy, and as science it is, from the point of view of today's knowledge, approximately 100 per cent wrong. Its interest lies in its concise and judicious summarizing of current opinion and in its long life, which is good evidence of the persistence of old ideas during a period

---

[14] John Clarke, *Physical Science in the Time of Nero, being a Translation of the "Quaestiones naturales" of Seneca* (London, 1940), p. li. Sir Archibald Geikie, the geologist, contributed to this volume thirty-four pages of notes which are of great value as a scientist's evaluation of Seneca's speculations.

[15] Because of its copious notes I have found the translation by John Bostock and H. T. Riley (6 v., London, 1855-1857) especially useful. The edition by H. Rackham (Loeb Classical Library, 10 v., Cambridge, Mass., 1938-1952) is more readily accessible.

[16] Paul H. Kocher, *Science and Religion in Elizabethan England* (San Marino, Calif., 1953), p. 135.

of far-reaching change—the period, indeed, of the scientific revolution.

The text of 1563 is reproduced below without change except for the substitution of the short for the long *s*, the expansion of contractions, and the correction of obvious misprints, most of them also corrected in the later printings. Departures from the copy-text are specified in the footnotes except for the correction of turned letters, which is performed silently. Roman type has been substituted for the black letter of the original. The marginal notes of 1563 have been omitted because, with rare exceptions, they merely repeat a word or phrase of the text.

[$\pi 3^r$] TO THE RIGHT *HONORABLE THE LORDE* Robert Dudley,[1] Maister of the Quenes maiesties horse, Knight of the most Noble order of the garter, and one of the Quenes maiesties priurie Counsell. William Fulce, wisheth increase of grace and heauenly giftes, in perfect health and true honor, long to continue.

[1] Fulke's many favors from Dudley (1532-1588), created earl of Leicester in 1564, have been summarized in the introduction. *A Goodly Gallerye* is the eighth title in Eleanor Rosenberg's list of ninety-four books dedicated to Leicester (*Leicester, Patron of Letters,* New York, 1955, p. 356).

THE moost myghtye Monarchie[2] of the *Romain* Empire *Octauius Augustus,* (ryghte honourable Lord) did shewe so great liberalitie, or rather magnificence towardes all them that bare him good will, that he also most largely rewarded sondry persones that had tought Popingayes and other birdes, to pronounce some salutation (as he passed by them) in his prayse and commendation. At the length a poore yonge man, allured (as it seemed) by his bountifull remuneration of other: had tought a Crowe (an vntowardly[3] byrde) after the same manner to salute themperour. Who percei[$\pi 3^v$]uing his purpose, that it was rather for hope of gayne, then testimony of good wyll: answered that he had store inough already, of suche saluters at home, meaning those Popingayes and other birdes, which at the first for their strangenes, he had dearly payed for.[4] In lyke manner (right honorable Lorde) when I present the salutation of this myne vntowardly byrde, a Crowe in comparison of suche pleasaunt popingayes, as haue been heretofore offered vnto your honorable Lordship: you may haue iust cause of suspition, that being moued with your former liberalitie and magnificence towards other, I doe as the Poet sayeth:

Occultum cautus decurrere piscis ad hamum.[5]

---

[2] Monarchie] 1571, 1563 Monaychie. Error for *Monarche?*

[3] vntowardly] awkward, unteachable.

[4] This anecdote is related by Macrobius (*Saturnalia* ii. 4. 29-30). It has a happy ending which Fulke omits as foreign to his purpose.

[5] *hamum*] Horace, *Ep.* I. vii. 74. In most editions this line reads "Occultum visus decurrere piscis ad hamum" ("seen to run like a fish to the hidden hook," tr. H. R. Fairclough, Loeb Classical Library). Perhaps this line has been confused with another in a later epistle (I. xvi. 50-51), "cautus enim metuit . . . opertum miluus hamum."

And so if your gentle[6] nature and noble disposition could suffer, as a crauer of vndeserued benefites, to reiect both me and my present. But sith[7] my state is such, that I can not testifie the good wyll of myne heart towardes your honour, otherwyse then by such meanes as this: I was bolde to referre that suspition to the iudgement of your wysdome and humanitie (knowing my selfe cleare from suche intent) and to commit this vnbew[$\pi 4^r$]tifull byrd, vnder the wynges of youre honorable protection, trusting that the same (whiche I take as a commen defence of all good learning) shall not only at this time be my stay and refuge, but also hereafter to more worthy attempts a continual encouragement. But specially at this tyme, I was bolde to enterprise the matter, for that one *Iames Rowbothum,* a man of notable impudens (that I saye no worse of him)[8] abusinge your singular humanitie and gentlenes expressing thexample of one *Bathillus,*[9] or rather (that I may continue in the allegorie of birdes) of *Esopes* crowe,[10] hath not ben ashamed to dedicate vnto your Lordship of late a treatise of myne, which I gathered out of diuerse writers, concerning the *Philosophers game:*[11] notwithstanding he was streightly commaunded to the contrary by the right honorable and reuerent father, my Lord of London,[12] of whome also I was exhorted and encouraged to dedicate the same

[6] gentle] generous, noble.

[7] sith] since.

[8] him)] 1563 him $\wedge$.

[9] *Bathillus*] According to Donatus, Bathyllus, a composer of pantomimes, claimed authorship of a distich in praise of Augustus which Vergil had composed. See note 14 below.

[10] *Esopes* crowe] Conceivably the crow on whose vanity the fox played in order to steal his meat. The crow who adorned himself with other birds' plumage is alluded to by Horace (*Ep.* I. iii. 18-20), not Aesop, but in the sixteenth century any beast fable might be associated with Aesop.

[11] See introduction, n. 6. Both Fulke and Ralph Lever (†1585) accused Rowbotham of mangling their work.

[12] my lord of London] the Bishop of London, Edmund Grindal (1519?-1583), afterward (1575-1583) Archbishop of Canterbury.

vnto your honour, my selfe. Whiche though nowe through his importunitie and disobedience, it be intercepted, and the booke defaced with his rude rythmes and peuish[13] verses: [π4^{v}] yet I thought best to geue your Lordeship vnderstanding that your honorable protection which is and should be the defence of learning and learned men, might no more be a boldening to such ignorant and vnhonest persones. And likewyse most humbly to desire your honour that though myne Epistle dedicatory, be not annexed to that booke, yet considering by what pertinacitie I was hindered, you would accept that booke also together with this, as an homely present of myne, of which I may conclude as one did in like case.

Illum ego composui librum, tulit alter honorem
Sic vos non vobis lacte tumetis oues.
Sic vos non vobis præda agit ampla canes
Sic vos non vobis conditis antra feræ,
Sic vos non vobis pondera fertis equi.[14]

Thus sparing to trouble your honour any longer, either with complaint or excuse, I desyre almightie God to multiplie his blessinges toward you, that abounding in all good giftes bothe bodely and ghostly, you may haue long life in health and honour, to his glory, the profit of other, and your endles comfort. Amen.

[13] peuish] silly, senseless (*New English Dictionary,* Peevish *a.* 1).

[14] Vergil confounded the plagiarist Bathyllus (n. 10) by composing these verses on a theme set by the emperor, "Sic vos non vobis," when the latter was unable to do so. The text quoted by Fulke is quite different from the traditional one.

[A1r] ❡ The First Booke.

For as muche as we entend in this treatise, to declare the causes of all those bodyes, that are generatede in the earthe, called *Fossilia,* as well as thos other *Impressiones,*[1] named of ther height, *Meteora* (which no wryter hetherto hath done, that we haue sene) the comen definicion gyuen by most wryters, in no wyse wyll serue vs, and whether we maye borowe the name of *meteoron* to comprehende the whole subiect of oure woorke we are not all together out of doubt, all thoughe the philosopher,[2] deryuinge it from doutfullnes, geueth vs som collore so to take it,[3] and peraduenture we myght be as wele excused to aplye it to *mineralls,* as other auters ar to vse it for earthequakes: yet to auoyde all occasions of cauellynge at wordes, we shall bothe defyne and also describe the subiect of oure matter on thys manner: yt is a body compounde with out lyfe naturalle: and yet to stoppe one hole, because heare wanteth the name of the thinge to be deffinede, yt is no newe thinge, to theim that haue redde *Aristoteles* workes to fynde a diffinition, of that whereof ther [A1v] is no name. But what nede you be so precise (wil some man say) mean you so to proceade in all your discourse? no verely, but because many of quicke iudgement not considering the stile to be attempered, to the capacitie of the Readers, will impute the

---

[1] *Impressiones*] atmospheric influences, conditions, or phenomena (*New English Dictionary,* Impression *sb.* 6).

[2] the philosopher] Aristotle.

[3] so to take it] Apparently Fulke refers to the statement at the beginning of the *Meteorologica* (338b): "It [meteorology] studies also all the affections we may call common to air and water, and the kinds and parts of the earth and the affections of its parts. These throw light on the causes of winds and earthquakes and all the consequences the motions of these kinds and parts involve. Of these things some puzzle us, while others admit of explanation in some degree." (Tr. E. W. Webster, *The Works of Aristotle,* ed. W. D. Ross, 3, 1931.)

plaines to the ignoraunce of the authors, we thought good in the beginning to pluck that opinion out of their mindes (that as the common saying is) they may know we haue skill of good maners though we litle vse them.

These meteors are deuided after thre maner of ways,[4] first into bodies perfectly and imperfectly mixed. Secondly into moist impressions and drie. Thirdly into fiery, aery, watery and earthly. According to this last diuision, we shall speake of them in fowre bookes followyng. But first, we must be occupied a litle in the general description of the same, that afterward shalbe particularly intreated of.

¶ *Why they be called vnperfectly mixed.*

THey are called vnperfectly mixed, because they are very sone chaunged into another thing, and resolued into their proper elementes of whiche they do moste consist, as do all impressions, fyrie, ayrie, watrie, as snowe into water, cloudes into [A2r] waters, etc.

¶ *Why they be called perfectly mixed.*

The last sort namely earthly, *Meteores* are called perfectly mixed, because they wil not easely be chaunged and resolued from that forme which they are in, as be stones, metalles and other mineralles.

According to the qualitie of the matter, they are diuided into moist and drie impressions, consisting either of *vapores* or *exhalations. vapores* are called moist, and *exhalations* drie, whiche termes must be well noted, because they must be much vsed.

---

[4] thre maner of ways] Fulke's organization is scholastic in tone, but I have found no precise origin for it. That of Jerome Cardan (*De subtilitate,* 1551; *De rerum varietate,* 1557) is very similar. This method of classification, obviously useful for pedagogy, survived till the last decades of the seventeenth century, e.g., in Charles Morton's *Compendium physicae* (ed. Theodore Hornberger, Publications of the Colonial Society of Massachusetts 33, Boston, 1940; see p. 7).

¶ *Of the general cause of al Meteores and first of the materiall cause.*

The mater whereof the moste part of *Meteores* dooth consiste, is either water or earth, for out of the water, proceade vapors, and out of the earth come exhalations.

Vapor as the Philosopher sayeth,[5] is a certain watrie thing, and yet is not water, so exhalation hath a certain earthly nature in it, but yet it is not yearth.

For the better vnderstanding of vapors, vnderstande that they be as it were fumes or smokes, warme and moist, whiche will easely be resolued into water, muche like to the breath that proceadeth out of a [A2v] mans mouth, or out of a pot of water standing on the fiere. These vapors are drawen vp from the waters and watry places, by the heate of the Sunne, euen vnto the midle region of the ayre, and there after diuerse maner of meating with coldnes, many kynde of moist *Meteors* are generated, as sometime cloudes and rayne, sometime snowe and hayle, and that suche *vapors* are so drawen vp by the Sunne, it is playne by experience: for if there be a plash of water on a smothe and hard stoone, standing in the heate of the sunne, it wyl soone be drye, whiche is none otherwyse, but that the sonne draweth vp the water in thinne *vapores,* for no man is so fonde[6] to saye, that it can sinke into stoone or metall, and it is as great foly to thinke, it is consumed to nothyng: for it is a general rule, that that whiche is once a thing, can not by chaunging become nothyng, wherefore it followeth that the water on the stoone, as also on the earth, is for the moste part drawen vp, when the stoone or earth is dried.

Exhalations, are as smokes that be hoat and drie, whiche because they be thinne, and lyghter then *vapors,* passe the lowest and midle region of the ayre, and are caried vp

[5] as the Philosopher sayeth] The reference appears to be to Aristotle's *Meteorologica* 340b. Most of Fulke's first book amplifies the first chapters of the *Meteorologica* and *De generatione et corruptione*.

[6] fond] foolish.

euen to the highest region, wherefor the excessiue heat, by nearenes of the fier, they [A3r] are kindled, and cause many kinde of impressions. They are also sometimes *viscose,* that is to say clammy, by reason wherof, thei cleuing together and not being dispersed, are after diuerse soortes set on fier, and appeare somtims like Dragons, somtim like Goats somtime like candels somtime like speares.

By that which is spoken of *vapors* and *exhalations,* it is euident that out of the fier and theayer no matter wherof *meteores* shold consist can be drawen, because of their subtiltie and thinnes. For al *exhalation* is by making a grosser body more thinne, but the fier (we mean the elemental fier, and not the fire of the kitchen chimney) is so subtil and thin, that it cannot be made thinner: likewise the ayre is so thin, that if it be made thinner, it is changed into fire, and as the fire if it wer made thicker, wold becom aire, so the aire, being made grosser, wold be turned into water. Wherfor to conclude this part, the great quantitie of matter, that causeth these *meteores,* is taken out of the earth and the water. As for the aire and the fire, they ar mixed with this matter as with al other things, but not so abundauntly, that they may be sayd the material cause of any *Meteore,* though without them none can be generated. The efficient cause of all *Meteores* is that cause, whiche maketh them, euen [A3v] as the Carpenter is the efficient cause of an house. This cause is ether first or second.

The first and efficient cause is god, the worker of all wonders, accordinge to that testimony of the Psalmist, whiche sayeth. Fier, haile, snowe, yse, wynde and stormes,[7] do his

---

[7] stormes] Ps. cxlviii. 8. The 1563 text reads *stones,* an evident misprint, since on the title page, where the same verse is quoted, the word is *stormes,* and here the correction was made in the 1602 text. "Wind and storm" is the reading of the Bishops' Bible; the *Book of Common prayer* and the Geneva version read "stormy wind." No version reads "do his will and commandment" but instead "fulfilling his word" or "which execute his word."

will and commaundement, he sendeth snowe lyke woll,[8] etc. Almightie God therfore beinge the firste, principall and vniuersall cause efficiente of all naturall workes and effectes, is also the first cause of these effectes, whose profit is great, and operation maruelous.

The second cause efficient, is double, either remote that is to saye, farre of, or next of al. The farther cause of them as of all other natural effectes, are the same, the sonne, with the other planetes and sterres, and the very heauen it selfe in which they are moued. But chiefly, the Sunne by whose heate all or at lestwyse, the moste part of the vapors and exhalations are drawen vp.

The next causes efficient as the firste qualities, ar heate and colde, whiche cause diuerse effectes in vapors and exhalations, but to retorne to the heate of the Sunne, whiche is a very neare cause, it is for this [A4r] purpose two wayes considered.

One waye, as it is meane[9] and temperate

Otherwise, as it is vehement and burning

The meane, is by whiche he draweth vapors out of the water and exhalations out of the earth, and not onely draweth them out, but also lifteth them vp very high from the earth, into the ayer, where they ar torned into diuerse kinde of *Meteores*.

The burning heate of the Sunne is, by which he burneth dissipateth and consumeth the vapors, and exhalations before he draweth them vp, so that of them no *Meteors* can be generated.

These two heates, proceade from the Sunne either in respect of the place, or the tyme, but moste properly according to the casting of his beames either directly or vndirectly.

[8] like woll] Ps. cxlvii. 16.
[9] meane] average, middling.

In places where the Sunnes beames strike directly against the earth, and the water, the heat is so great, that it burneth vp the exhalations and vapors, so that there are no fiery *Meteors,* muche lesse watery as it is in the South partes of the world, vnder and neare to the Equinoctiall lyne.

But in places where the beames are cast indirectly, and obliquely, and that where [A4v] they are not to nyghe to the direct beames, nor to farre of from them: there is a moderate heate, drawyng out great aboundaunce of matter, so that in those contries, many *Meteores* of many sortes are generated,[10] as in the farre North partes are few, but watrie impressions. Also in *Autumne* and Sprynge, are oftener *Meteores* seen, then in Sommer and Wynter, except it be in such places, where the Sommer and Wynter are of the temper of Spryng and *Autumne.* Let this be sufficient, for the efficient causes of impressions, as well first and principall, as second and particular. Concerning the formall and finall cause, we haue litle to saye because the one is so secret, that it is knowen of no man, the other so euident that it is playne to all men. The essentiall forme of all substaunces. Gods wisdome comprehendeth, the vniuersall chiefe and last end of all thinges, is the glory of God. Mydle endes (if they may be so called) of these impressions are manifold profites, to Gods creatures, to make the earth fruitfull, to purge the ayre, to sett forth his power, to threathen his vengeaunce, to punyshe the worlde, to moue to repentaunce: all the which are referde to one end of Gods eternall glory, euer to be praysed. Amen.

[A5r] ¶ *Of the places, in whiche they are generated.*

THE places in whiche *Meteors* are caused, be either the ayre or the earth, in the aire be generated rayne, hayle, snow, dew, blasing starres, thonder, lightning etc. In the

[10] are generated] 1602; 1563 as generated.

earth be welles, springs, earthquakes, metalls, minerals, etc. made, and as it were in their mothers belly begotten and fashioned. But for the better vnderstanding hereof, such as haue not tasted the principles of *Philosophie,* must consider that ther be iiij. elements, Earth, water, Ayre, and Fire, one compassing another round about, sauing that the waters by Gods commaundement ar gathered into one place, that the land might apeare. The highest is the spere of the fire, which toucheth the hollownes of the Moones heauen, the next is the ayre, whiche is in the hallownes of the fyer, the ayre within his hollownes, comprehendeth the water and the earth, whiche bothe make but one *Spheare* or *Globe,* or as the commen sort may vnderstande it one Bal. So eche element is within another as the skales of a perle, ar on aboue another, or (to vse a grosse similitude) as the pieles[11] of an onion, are one within another, after the same sort from the highest heuen to the earth, that is lowest, one part that is greater compasseth [A5ᵛ] round about another that is lesser. But for this present purpose it is to be knowen, that the ayer is diuided into thre regions, the hyghest, the midle, and the lowest. The hyghest, because it is next to the region of the fier, is exceading hoate, the lowest beinge next the earth and the waters, is temperat, and by *repercussion* or striking back of the sunne beames waxeth hoate, and by absence of them is made colde, being subiect to Wynter and Sommer. The midle region of the ayre, is always exceading cold, partly because the sonne-beames, can not be cast back so highe, and partly because the cold that is there, betwene the heate aboue and the heate beneath it, is so kept in that it can not get out, so that it must nedes be excessiuely colde. For the water and the earthe being both colde elementes, after the sunnesetting in

---

[11] pieles] skin (this use antedates by twenty years the earliest example given by the *New English Dictionary*).

the nighte season dooe coole the aire, euen to the midle region. But in the morning the sunne rysing warmeth the ayre, so farre as his beames whiche are beaten back from the earth and the water, can extende and reache, whiche is not so highe as the midle region, and by heate on both sides, is inclosed and kept, sauing that a litle thereof falleth downe in the night, [A6r] which the next day with much more is driuen backe againe. Wherefore this region being so colde, is darke and cloudye, in so much that some doting Diuines haue imagined, purgatorie to be there in the mydle region of the ayre.

In the hyghest region, be generated Cometes or blasing starres, and suche lyke of diuerse sortes.

In the midle region cloudes, rayne, stormes, wyndes, etc. In the lowest region, dewe, frost, horefrost, mistes, bryght rods,[12] candels burning about graues, and gallowses, where ther is store of clamy fatty or oyly substaunce, also lightes and flammyng fiers, seene in fieldes, etc.

¶ And thus muche for the generall causes of all *Meteores.*

[A6v] ❧ The seconde Booke of fyery *Meteores.*

A Fiery impression, is an *exhalation* sett on fire, in the highest or lowest region of the ayre, or els apearing as though it were set on fire and burning.

They are therfore diuided into flames and aparitions. Flames are they, whiche burne in deade and are kindled with fire. These are discerned[13] by iiij. wayes, by the fashion of them, by their place, by the abundaunce of their matter, and by the wante of their matter. Their placing is after the aboundaunce and scarsetie of the matter wherof they consist, for if it be great, heauie and grosse, it cannot be caried so

---

[12] bryght rods] Translates the *bacula* or *virgae* of the Latin Aristotle (*Meteorologica* 341b, 372a, etc.).

[13] discerned] differentiated.

farre as the midle region of the aire, and therfore is set on fire in the lowest region, if it be not so great, light, and full of heat, it passeth the midle region and ascendeth to the highest, where it is easely kindled and set on fire.

According to their diuerse fashions, they haue diuers names, for they ar called, burning stoble,[14] torches, daunsing of leapinge Goates,[15] shooting or falling starres, or can-[A7r]dels, burning beames, round pillers, spears shieldes, globes or bowles, fierbrandes, lampes, flying dragons or fire-drakes, pointed pillors or broched[16] steples, or blasing stars, called *Cometes*. The time when these impressions doth most apere, is the night season, for if they were caused in the day time, thei cold not be seen, no more then the starres be seen, because the light of the sunne which is much greater, dimmeth the brightnes of them being lesser.

¶ *Of the generation of the impression, called burned stoble or sparcles of fire.*[17]

THE generation of this *Meteore* is this, whan the matter of the *Exhalation* is in all partes a lyke thynne, but not compacted or knit together, then some parte of it being caried vp into the highest region, by the fiery heate is set on fier before another part, that commeth vp after it, and so being kindled by lytle and lytle, it flyeth abrooode lyke sparkles out of a chymney, in so much that the common people suppose, that an infinit number of starres fal down where as it is nothing els, but the *Exhalation* that is thinne kindled in many partes, sparkling as when sawe dust or coole dust is cast into the fyre.

---

[14] burning stoble] Translates the *ardens stipula* of the Latin Aristotle (*Meteorologica* 341b).

[15] Goates] fiery meteors (this use antedates by ninety-three years the earliest example given by the *New English Dictionary,* Goat *sb.* 2c).

[16] broched] (of stone) indented with a chisel.

[17] spercles of fire] *Cf. Meteorologica* 341b-345a. Many of the names, as well as the general theory, are taken from Aristotle.

[A7v] ¶ *Of Torches.*

Torches of fyer brandes, are thus generated: when the matter of the exhalation is long and not broad, being kyndled at one end therof, in the highest region of the ayre, it burneth lyke a Torche or fyer brande, and so continueth, till all the matter be burned vp, and then goeth out, none otherwyse then a Torche when all the stuffe is spent must nedes burne no longer.

¶ *Of dansyng or leaping Goates.*

DAnsyng Goates, are caused when the exalation is diuided into twoo partes, as when twoo torches be seen together, and the flame appeareth to leape or daunce from one parte to the other, much lyke as balls of wylde fyer daunce vp and downe in the water.

¶ *Of shotyng and falling Starres.*

A Flying, shooting, or falling S[t]arre, is when the exhalation being gathered as it were on a round heape, and yet not throughly[18] compacted in the hyghest parts of the lowest region of the ayer, beynge kyndled, by the soden colde of the mydle region is beaten backe, and so appeareth as though a Starre should fall, or slyde from place to place. Sometyme it is generated [A8r] after another sort, for there is an exhalation long and narrowe, whiche being kyndled at one ende burneth swiftly, the fyer ronning from ende to ende, as when a silke thread is set on fyer at the one end. Some saye it is not so much set on fyer, as that it is direct vnder some Starre in the firmament, and so receiuing light

---

[18] throughly] thoroughly.

of that starre, semeth to our eyes to be a starre. In deade some times it may be so, but that is not so alwayes, nor yet most commenly, as it may be easely demonstrated. The *Epicurians*[19] as they are verye grosse in determining the chief goodnes, so they are very fonde in assigning the cause of this *Meteor.* For they say, that the starres fall out of the firmament, and that by the fall of them, both thonder and lyghtning are caused: for the lightening (say they) is nothyng els but the shyning of that starre that falleth, which falling into a watrie clowde, and being quenched in it, causeth that great thonder, euen as whoat yron maketh a noyce if it be cast into colde water. But it is euident that the starres of the firmament can not fall, for God hath set them fast for euer, he hath geuen them a commaundement whiche they shal not passe.[20] And though they shold [A8v] fall into the cloude, yet could they not rest there, but with their weyght being dryuen downe, would couer the whole yearth. For the least starre that is seen in the firmament, is greater then all the earth. Here wyl steppe foorth some mery fellow, which of his conscience[21] thinketh them not to bee aboue three yardes about, and saye it is a loude lye, for he can see within the compasse of a bushell[22] more then xx. starres. But if his bushell were on fyre xx. myle of, I demaunde how bygge it would seem vnto him? He that hath any wyt, wil easely perceiue, that starres being by al mennes confession, so many thousand myles distant from the earth, must neades be very great, that so farre of should be seen in any quantitie.

Thus muche for the shootying or fallynge Starres.

---

[19] *Epicurians*] Fulke who refers elsewhere to Epicurus (see p. 84), does not quote Lucretius's *De rerum natura,* a book of considerable relevance to his subject (*cf.* vi. 145-149). *Cf.* also Pliny, *Historia naturalis* ii. 43.

[20] passe] neglect, disregard (*New English Dictionary,* Pass *v.* 29).

[21] of his conscience] according to his firm conviction.

[22] bushell] bushel basket or other vessel.

¶ *Of burnyng Candels.*

WHen the *Exhalation* caried vp into the hyghest part of the ayre is in al partes thereof of equall and lyke thynnes, and also long, but not broade, it is set on fyre and blased lyke a candle, vntyll the *Exhalation* be quite consumed.

¶ *Of burning Beames and round Pillers.*

THese are caused, when the *Exhalation* [B1r] being long and not very broade, is sett on fyre, all at once and so burneth lyke a great beame or logge. The difference of beames and pyllers is this, for beames are when they seeme to lie in lengthe, in the ayre, but they are called pillers, when they stande right vp, the one end nearer to the earth, then the other.

¶ *Of burning Speares.*

BUrning spears are generated, when a great quantitie of *Exhalations,* which may be called a drie cloude, is set on fire in the myddest, and because the cloude is not so compacte, that it shoulde sodenly rende, as when thonder is caused, the fyre breaketh out, at the edges of the clowde, kendlyng the thynne *Exhalations,* which shoot out in great nomber lyke fyry spears, or darts, longe and very small, wherfore they continue not long, but when they fayle, within a short whyle after, more fyre breakynge out, they shoot out as many more in their place, and lykewyse, when they ar gone, other succeade, if the quantitie of the matter wyll suffise, more then a doosen courses. This impression was seen in London, *Anno dom.* 1560. the thirty daye of [B1v] *Ianuary,* at eight of the clocke at night, the ayer in all other places beyng very darke, but in the North east where this cloude burned, it was as lyght as when the daye breaketh, towarde the Sunne rysyng, in so much, that playne shadow of thyngs opposite, was seen. The edge of this cloude was in fashion lyke the Raynbowe, but in collour very bryght, and often tymes casting foorth almoste innumerable dartes, of wonderful length lyke squybbes, that are cast vp into the ayre, sauing that they moued more swiftly then any squybbes.

¶ *Of shieldes, Globes or Bowles.*

THese *Meteores* also haue their name of their fashion, because they are broad, and appeare to be rounde, otherwyse their generation differeth not from the cause of the lyke impressions before mentioned.

¶ *Of Lampes.*

THe lampe consisteth of an *Exhalation,* that is broade and thick, but not equally extended, namely smaller at one end then at another, which being kindled about the middest therof, burneth like a lampe. The cause why, as wel this impression, as many other, apeareth round, is not for that alwayes they ar round in deede, but because [B2r] the great distance causeth them to seem so, For euen square formes, farr of seem to be round. It is written, that a lampe fel down at Rome when *Germanicus Cesar,* set forth the sight of sword players.[23]

¶ *Of flying Dragons or fyre Drakes.*[24]

FLying Dragons, or as Englyshmen call the fire drakes, be caused on this maner. When a certen quantitie of *vapors* ar gathered on a heape, being very near compact, and as it wer hard tempered together this lompe of *vapors*

---

[23] sword players] This story is mentioned both by Seneca (*Quaestiones naturales* i. 1) and by Pliny (*Historia naturalis* ii. 25).

[24] fire drakes] Fulke's use of *firedrake* in this sense is the earliest recorded by the *New English Dictionary.*

assending to the region of cold, is forcibly beaten backe, whiche violence of mouing, is sufficient to kindle it, (although som men will haue it to be caused betwene ij. cloudes a whote and a cold) then the highest part, which was climming vpward, being by reason more subtil and thin, apeareth as the Dragons neck, smoking, for that it was lately in the repuls bowed or made crooked, to represent the dragons bely. The last part by the same repulse, turned vpward, maketh the tayle, both apearing smaller, for that it is farther of, and also, for that the cold bindeth it. This dragon thus being caused, flyeth along in the ayre, and somtime turneth to and fro, if it meat with a cold cloud to beat it back, to the great [B2^v] terror, of them that beholde it, of whom some called it a fyre Drake, some saye it is the Deuill hym selfe, and so make report to other. More then sixtene yeares ago,[25] on May daye, when many younge folke went abroade early in the mornyng, I remember, by sixe of the clocke in the forenoone, there was newes come to London, that the Deuill the same mornynge, was seene flyinge ouer the Temmes: afterward came worde, that he lyghted at Stratforde, and ther was taken and sett in the stockes, and that though he would fayne haue dissembled the matter, by turning hym selfe into the likenes of a man, yet was he knowen welinough by his clouen feet. I knowe some yet alyue, that went to see hym, and returning affirmed, that he was in deed seen flyng in the ayre, but was not taken prysoner. I remember also that som wyshed he had been shoot at[26] with gons, or shaftes as he flewe ouer the Temmes. Thus do ignorant men iudge of these thynges that they knowe not, as for this Deuill, I suppose it was a flyinge Dragon, whereof we speake, very fearefull to loke vpon, as though he had life, [B3^r] because he moueth, where as he is nothing els but cloudes and smoke, so mightie is

---

[25] more then sixtene yeares ago] 1563; 1602 More than 47. yeares agoe.
[26] shoot at] shot at.

God, that he can feare[27] his enemies, with these and suche lyke operations, whereof some examples may be founde in holy scripture.

¶ *Of the Pyramidall pyller lyke a spire. or broched steeple.*

THis sharpe poynted piller, is generated in the hyghest region of the ayre, and after this sort. When the *Exhalation* hath much earthly matter in it, the lighter partes and thinner (as their nature is) ascending vpwarde, the grosser, heauier, and thycker, abyde together in the bottome, and so is it of fashion great beneath, and small pointed aboue, and being set on fire it is so seen, and thereof hath his name.

¶ *Of Fyre scattered in the ayre.*

FYre scattered in the ayre, or illuminations, are generated in the lowest region of the ayre, when very drye and whote *Exhalations,* are drawen vp and meeting with colde cloudes, are sent back agayne, which motions setteth them a fyre, whose partes, being not equally thycke or ioyned together, seeme as [B3v] though fyer were scattered in the ayre. Yea sometimes, the whole ayre seemeth to burne, as though it would raine fyre from heauen, and so it hath come to passe, burning both cities and townes. Then iudge, how easy it was for God to raine fire vpon Sodome and Gomor, for their sinnes and wickednes.

¶ *Of lights that goeth before men, and followeth them abrode in the fields by the night season.*

THere is also a kind of light, that is seen in the night season, and seemeth to goe before men, or to followe them, leading them out of their waye vnto waters, and other

[27] feare] frighten.

daungerous places. It is also very often seen in the night, of them that sayle in the Sea, and sometyme will cleaue to the mast of the shypp, or other highe partes, somtyme slyde round about the shyppe, and either rest in one part till it go out, or els be quenched in the water. This impression seen on the lande, is called in latin, *Ignis fatuus,* foolish fyre, that hurteth not, but only feareth foules. That whiche is seen on the Sea, if it be but one, is named *Helena,* if it be twoo, it is called *Castor* and *Pollux.*

The foulische fyre, is an *Exhalation* [B4r] kendled by meanes of violent mouing, when by cold of the night, in the lowest region of the ayre, it is beaten downe, and then commonly, if it be light, seeketh to ascende vpward, and is sent down againe, so it danseth vp and downe, Els if it moue not vp and downe, it is a great lompe of glueysh or oyly matter, that by mouing of the heate in it selfe, is enflamed of it selfe, as moyst haye wyll be kyndled of it selfe. In whote and fenny countries, these lyghtes are often seen, and where as[28] is abondaunce of such unctuus and fat matter, as about churchyardes wher through the corruption of the bodies ther buried, the earth is ful of suche substance, wherfore in churchyardes, or places of common buriall, oftentimes ar such lightes seen, which ignorant and superstitious fooles, haue thought to be soules tormented in the fyre of purgatorie. In dede the deuill hath vsed these lightes (although they be naturally caused) as strong delusions to captiue the myndes of men, with feare of the Popes purgatorie, wherby he did open injury to the bloud of Christ, which only purgeth vs from al our sinnes and deliuereth vs from al torments, both [B4v] temporall and eternal, according to the saying of the wyse man, the soules of the ryghteous are in the hands of God, and no torment toucheth them. But to returne to the lightes in

[28] where as] wherever.

whiche, there ar yet twoo thinges to be considered. First, why they leade men out of their waye. And secondly, why they seeme to follow men and go before them. The cause why they leade men out of the waye, is, that men whyle they take hede to such lights, and are also sore afrayde, they forgett their waye, and then being ones but a litle out of their waye, they wander they woote not whether, to waters, pyttes, and other very daungerous places. Which, when at lengthe they happe[29] the waye home, wyll tell a greate tale, how they haue been lead about by a spirite in the lykenes of fyre. Nowe the cause why they seeme to goe before men, or to followe them, some men haue sayde to be the mouing of the ayer by the goyng of the man, which ayre moued, shold driue them forward if they were before, and drawe them after, if they were behynd. But this is no reason at all, that the fire which is oftentimes, thre or fowre miles [B5^{r}] distaunt from the man that walketh, shold be moued to and froo by that ayre which is moued through his walkinge, but rather the mouing of the ayre and the mans eyes, causeth the fyre to seeme as though it moued, as the Moone to chyldren seemeth, if they are before it, to run after them: if she be before them, to run before them, that they can not ouertake her, though she seeme to be very neare them. Wherfore these lyghtes rather seeme to moue, then that they be moued in deade.

¶ *Of Helena, Castor and Pollux.*

WHen the lyke substaunce in the lowest region of the ayre, ouer the Sea by the lyke occasion is set on fyre, if it be one onely, it is called *Helena,* if ther be two, they ar called *Castor* and *Pollux*.[30] These impressions will oftentimes cleue

[29] happe] find by chance (but the *New English Dictionary,* Hap *v.*[1], gives no example of the transitive use).

[30] *Pollux*] Pliny ii. 26.

to the maste and other partes of the ships, by reason of the clammynes and fatnes of the matter. *Helena* was of the Heathen men, taken as a Goddesse the daughter of *Iupiter* and *Leda*. *Castor* and *Pollux,* were her brethren. *Helena* was the occasion that Troy was destroyed, therfore the Mariners by experience tryinge[31] that one [B5v] flame of fyre apearyng alone, signified tempest at hand, supposed the same flame to be the goddesse *Helena,* of whom they looked for nothing but destruction. But when two lightes ar seen together, they ar a token of fayre wether, and good luck, the Mariners therfor beleued, that they were *Castor* and *Pollux,* whiche sayling to seeke their syster *Helena,* beyng caried to Troye by *Paris,* were neuer seen after, and thought to be translated into the nomber of the Gods that gyue good successe to them that sayle, as we reade in the last chapter of the Actes of the Apostles, that the shyppe wherein S. Paule sayled, had a badge of *Castor* and *Pollux.*[32] A natural cause why thei may thus fore shewe either tempest or calmnes, is this. One flame alone may geue warning of a tempest, because that as the matter therof is compact, and not dissolued, so it is lyke, that the matter of tempeste (whiche neuer wanteth) as wynde and cloudes, is styll together, and not dissipated, then is it lyke not long after to aryse. By two flames together, may be gathered, that as this *Exhalation* whiche is very thycke is diuided, so the thycke [B6r] matter of tempest is dissolued and scattered abroade by the same cause that this is diuided. Therfore not without a reason, the Mariner to his mates may promyse a prosperous course.

---

[31] tryinge] finding out (*New English Dictionary,* Try *v.* 13).

[32] *Pollux*] Acts xxviii. 11.

ℂ *Of flames that apeare vpon the heares of men or beastes.*

THere is yet another kynde of fyry impression, which is flames of fire vpon the hears of men and beastes, especially horses. These are somtime clammy *Exhalations,* scatered abroade in the ayre in small partes, which in the night by resistaunce of the colde, are kendled, cleauyng on horses eares, on mens heades and shoulders that ryde or walke. In that they cleaue vpon heares, it is by the same reason, that the dewe wyll be seen also vpon heares or garmentes, whose woll is hyghe, as fryese[33] mantels and suche lyke. Another sorte of these flames, are caused, when mens or beasts bodies being chaffed, send forth a fat and clammy swet, which is in like maner kindled as the sparkes of fire that ar seen when a black horse is curried.[34] *Liuius* reporteth[35] of *Seruius Tullius,* that as he lay aslepe, being a childe, his heare seemed to be all on a flame, which [B6v] for all that did not burne his heare, or hurt him. The lyke historie he reciteth of one *Marius,* a Knyght of Rome, that as he made an oration to his Souldiors in Spaine, they sawe his head burning on a lyght fyre, and he hym selfe not ware of it. Thus muche concerning these flames.

¶ *Of Comets or blasing Starres.*

A *Comete* is an *Exhalation,* whote and drye, of great quantitie, fat and clammye, harde compacte lyke a greate lompe of pitche, which by the heate of the sunne, is drawen out of the earth, into the hyghest region of the ayre, and there by the excessiue heat of the place, is set on

[33] fryese] coarse woolen cloth with a nap.
[34] curried] 1602 curryed; 1563 curred.
[35] *Livius* reporteth] i. 39. The "Marius" referred to below is Lucius Marcius (xxv. 39).

fire, apearing lyke a starre with a blasinge tayle, and sometyme is moued after the motion of the ayre, whiche is circuler, but it neuer goeth downe out of the compasse of syght, though[36] it be not seen in the daye tyme for the bryghtnes of the sunne, but styll burneth vntyll all the matter be consumed. An argument of the greatnes is this, that there was neuer any *Comet* yet perceyued, but at the lest it endured seuen dayes, but much longer they haue bene seen, namely fowrtye [B7r] dayes long, yea, lxxx. dayes, and some syxe moneths together.[37] Wherfore, it must neades be a wonderfull deale of matter, that can gyue so much noryshement, for so great and feruent fyre, and for so long a tyme. There are consydered in a *Comet*, specially the collor and fashion, which both aryse of the disposition of the matter. Their collours, be either whyte, ruddye, or blewe. If the matter be thynne, the collour is whyte. If it be meanly[38] thycke, then is the *Comet* ruddy, after the collour of our fyre, but when the matter is very thycke, it is blewe, lyke the burning of Brymstone. And as the matter is more and lesse, after this disposition: so is the *Comet* of collour, more or lesse lyk to these three principal collours, some yelowyshe, some duskish, some grenishe, some watchet[39] etc.

In fashion ar noted three differences, for eyther they seeme round, with beames round about, or with a beard hangyng downward, or els with a tayle stretched out sydelong, in lengthe. The first fashion, is when the matter is thickest in the myddest, and thinne rounde about the edges. The seconde is, when the *Exhala*-[B7v]*tion* is vpwarde thicke, and in lengthe downewarde also, meanly thycke. The third forme is lyke the seconde, sauynge that the tayle

---

[36] though] 1602; 1563 thought.
[37] syxe moneths together] *Cf.* Pliny ii. 22.
[38] meanly] moderately.
[39] watchet] light blue.

hangeth not downe, but lyeth asyde, and is commonly longer then the beard. The tyme of their generation is oftenest in *Autumne* or haruest. For in the spring, there is to muche moysture, and to lytle heate, to gather a *Comet.* In sommer, is so muche heate, whiche will disperse and consume the matter, that it can not be ioyned together. As for wynter, it is cleane contrary to the nature of a *Comet,* which is whoat and drye, wynter being cold and moist, therfore no tyme so meet as *Autumne.*

Now for so muche as many learned men, haue gone about to declare the signification of blasing starres, we will omit nothing that hath any shadowe of reason, but declare what is wrytten of them.

Such things as are set forth of the betokening of *Comets,* ar of two sortes: the first is of naturall, the second of ciuile or politike effects. They ar sayd to betoken drought, barynes[40] of the earth and pestilence.

Drought, because a *Comet* can not be generated without great heat, and muche [B8r] moisture is consumed in the burning of it.

Barrines, because the fatnes of the earth, is drawen vp, whereof the *Comet* consisteth.

Pestilence, forsomuch as this kynd of *Exhalations,* corrupteth the ayre, whiche infecteth the bodies of men and beastes.

The second sort, might wel be omitted, sauing that *Aristotle* him selfe, disdayneth not to seeke out causes for some of them. Generally it is noted of all Historiographers, that after the apearing of *Comets,* moste comenly followed, great and notable calamities. Beside this, they betoken (sayth some) warres seditions, changes of commen wealths and the death of Princes and noble men.

---

[40] barynes] barrenness.

For what times *Comets* do shyne, ther be many whote and drie *Exhalations* in the aire, which in drie men kindle heat, wherby they ar prouoked to anger, of anger commeth brawling, of brawling fighting and war, of warre victory, of victory chaunge of common welths. Then also Princes, liuing more delicatly then other men, ar more subiect to infection, therefor dye sooner then other men. If it wer lawful to reason of this sort, we might enduce them to betoken, not only these few things, but al other things that chanse in the worlde.

[B8v] Yet these predictions haue a shew of reason though it be nothing necessary: but it is a world to see,[41] how the *Astrologians,* dote in suche deuises. They ar not ashamed, to an earthly substaunce, to ascribe an heauenly influence, and in order of iudgement to vse them as verye starres, suerly by as good reason as to the celestiall starres, they atribute diuine influences and effectes. But this their fooly, hath been sufficiently detected by dyuerse godly and learned men, and this place requireth no longe discourse thereof. Wherfore this shall suffice, both for the naturall causes of blasyng starres, and also, for all flames in generall. It followeth therfore, that with like breuitie we declare the causes of fyery aperitions.

¶ *Of Aparitions.*

AN *Aparition,* is an *Exhalation* in the lowest or hyghest region of the ayre, not verely burning, but by refraction of lyght, either of the sunne or the Moone, seemeth as though it burned. Whiche appearaunce of collour, ryseth not of the mixtion[42] of the fowre qualities, as it doth in bodies perfectly mixed, as herbes, [C1r] stones etc. But only of the falling of light vpon shadowe. The light is in steade

---

[41] it is a world to see] it is wonderful to see.
[42] mixtion] mixture.

of whyte, and the shadowe or darkenes in steade of black. These diuersely mixed according to the diuerse disposition of the *exhalation,* which ministreth varietie by thicknes or thynnes, cause diuerse collours.

There be commonly recited three kindes or fyery apparitions.

¶ *Collours, wyde gapinges, and deepe hooles, whiche apeare in the cloudes.*

¶ *Of collours.*

Collours are heare ment, when there is nothing els to be noted, but the collours of the cloudes, and they are caused (as it is sayde) by casting the lyght into the shadowye cloude, accordyng as it exceadeth more or lesse in thicknes. wherof some be very bright whyte, and that is when the *Exhalation* is very thynne; some yealowish, when the *Exhalations* is thicker, somme ruddy, when it is meanly thicke. and very black when it is very thicke. The redd and ruddy collours are seen, only in the mornyng and euening, when the lyght of the sunne is not in his full force, for at other tymes of the daye, his lyght is to vehement, cleare, [C1v] strong, and pearsing. This much of collours.

¶ *Of wyde gaping.*

Wyde gaping is caused, when an *Exhalation* is thick in the middest, and thinne on the edges, then the light being receiued into it, causeth it to appeare as though the skye did rende, and fire breake out of it.

¶ *Of round opening Hiatus.*

These holes called *Hiatus,* differ from wyde gapinges in nothing, but that they be lessc, and therfore seeme as though they were depe pittes, or holes, and not rending or gaping, and these be those apparitions, that apeare fyery and yet bee not so in deade. Therfore let this be sufficient to haue shewed the natural causes of all fiery *Meteores.*

❧ The thirde Booke of aery impressions.

VNder the name of aery impressions, be comprehended, such *Meteores,* whose matter is most of the aire. Of this sort be windes, earth quakes, thonder, lightnings storme wyndes, whirlewyndes, circles, raynbowes, the white circle, called of som watling strete, many sunnes many mones.

¶ *Of Wyndes.* [C2r]

THe wynd is an *Exhalation* whote and drie, drawne vp into the aire by the power of the sunne, and by reason of the wayght therof being driuen down, is laterally or sidelongs caried about the earth, and this diffinition is to be vnderstanded, of generall wyndes, that blowe ouer al the earth, or els som great regions, but beside these, there be particular wyndes, whiche are knowen but only in som countries, and then not very large, these wyndes oftentimes haue another maner of generation. And that is on this maner. It must needes be confessed, that within the globe of the earth, be wonderful great holes, caues, or dongeons, in which when ayer abondeth (as it may by diuerse causes) this ayer, that cannot abide to be pinned in, findeth a litle hole in or about those countries, as it weare a mouth to break out of: and by this meanes, bloweth vehemently, yet the force and vehemens extendeth not far, but as the wynde that commeth forth of bellowes, neare the comming foorthe is stronge, but farre of, is not perceiued: So this particular wynd, in that countrye, where it breaketh forth, is very violent and strong, in somuch, that, it ouerthroweth both trees, and houses, [C2v] yet in other countries, not very farre distant, no part of that boisteous blast is felt. Wherfore this wynde differeth from the generall wyndes, both in qualities and substaunce or matter, for the matter of them is an *Exhalation,* and the qualities suche as the nature of the *Exhalation* is, very ayery, but not ayere in deade: but of this particular wynde, the matter and substaunce is moste commonly ayer.

There is yet a thyrde kynde of wynde, whiche is but a softe gentle and coole mouing of the ayre, and commeth from no certaine place (as the generall wynd doth) yea it is felt in the shadowe vnder trees, when in the whote lygth and shining of the sunne, it is not perceiued. It commeth whisking sodenly, very pleasaunt in the heate of the sommer, and ceaseth by and by. This properly is no wynde but a mouing of the ayre by som occasion.[43] As for the generall wyndes, thei blowe out of diuerse quarters of the ayre, nowe East, nowe West, nowe South, nowe North, or els mclininge to one of the same quarters. Amonge whiche the East wynde followyng the nature of the fyre, is whote and drie, the [C3r] South wynde expressing the qualitie of the ayre, is whote and moyste, the Westerne blast, agreing with the waters propertie is colde and moyst. The Northe that neuer was warmed with the heat of the sunne, being cold and drye, partaketh the conditions of the earth. The midle wyndes haue midle and mixed qualities after the nature, of those fowre principall wyndes more or lesse, as they encline toward them more or lesse.[44]

[43] som occasion] *Cf.* Bartholomaeus Anglicus, *De proprietatibus rerum* xi. 16; G. G. Pontano, *Meteororum liber* (p. 113 in the Aldine edition of 1533).

[44] more or lesse] "Fulke skilfully combined the three independent and opposing theses of Aristotle [*Meteorologica* 361a, that wind is a body of dry exhalations moving about the earth], Pliny [ii. 44, that winds are bred in caves], and Hippocrates [*Meteorologica* 348b, that wind is simply a moving current of air] into a single system which explained all types of wind" (S. K. Heninger, *Handbook of Renaissance Meteorology,* 1960, p. 108).

Generally the profit of all wyndes,[45] by the wonderfull wysdome of the eternall God, is wonderfull great, vnto his creatures. For besydes that these wyndes, alter the weather, some of them bryngyng rayne, some drynes, some frost and snowe, whiche all are necessary, ther is yet an vniuersall comoditie, that ryseth by the only mouyng of the ayre. Which were it not continually styrred, as it is, would soone putrifie, and beyng putryfied, would be a deadly infection to all that hath breath vpon the earth. Wherefore this wynde whose sounde we heare, and knowe not from whence it commeth nor whether it goeth (for who can affirme from whence it was raysed, or where it [C3v] is layde downe) as al other creatures besyde doth teach vs, the wonderfull and wyse prouidence of God, that we maye worthely crie out, with the Psalmist, and saye: O Lorde, howe manyfolde are thy workes,[46] in wysdome hast thou made them all, etc. Let this be sufficient, to haue shewed the generation of the wyndes.

¶ *Of earthquakes.*

AN earthquake, is a shaking of the earth whiche is caused by meanes of wynde and *Exhalations,* that be enclosed, with in the caues of the earth, and can fynde no passage, to breake foorth, or els so narrowe a waye that it can not be soone enoughe delyuered. Wherefore, with great force, and violence it breaketh out, and one whyle shaketh the earth, another whyle rendeth and cleaueth the same, sometyme it casteth vp the earth, a great heygth into the ayre, and some tyme it causeth the same, to synke a great depth downe, swallowyng both cyties, and townes, yea and also mightie great mountaignes, leauing in the place wher they stoode,

[45] of all windes] Fulke's argument from design is closely parallel to that of Seneca (v. 18).

[46] workes] 1602 works; 1563 wordes.

nothyng but great holes of an vnknowen depthe, or els great lakes [C4r] of waters.

¶ *Of diuers kindes of earthquakes.*

DYuerse authors wryte dyuerselye, of the kindes of earthquakes, some makyng more and some lesse, but we shall be content at this tyme to comprehende them in fowre sortes.

The first kynde is when the earth is shaken laterally, to one syde, whiche is when the whole force of the wynde dryueth to one place, and there is no other contrary motion to let it.[47] This wynde if it be not great shaketh the earth, that it trembleth as a man that hath a fyt of an agewe, and dothe no more harme, but if it be great and violent, it louseth the foundations of all byldinges,[48] be they neuer so stronge, and ouerthroweth whole cyties, but specially the great buildynges, and not onely buyldinges, but some tyme also casteth downe greate hylles, that couer and ouerwhelme, all the valley vnder them. Many noble and great cities, haue been ouerthrowen by this kynde of earthquake. It is wrytten that twelue of the mooste bewtifull cyties, and moste sumptuous buildyngs in all *Asia,* were ouerthrowne and vtterly [C4v] destroyed with an[49] earthquake.[50] Howe often, *Antiochia,*[51] yea within short tyme, was destroyed, they whiche haue redde the histories, can testifie. Howe terrible was the earthquake, that shooke *Constantinople*[52] a

[47] let] hinder.

[48] byldinges] 1602; 1563 bydinges.

[49] an] 1602; 1563 any.

[50] earthquake] Pliny (ii. 86) mentions the twelve cities.

[51] *Antiochia*] Robert Mallet, *Catalogue of Recorded Earthquakes from 1606 B.C. to A.D. 1850* (Report of the Twenty-second Meeting of the British Association for the Advancement of Science, 1853), pp. 5 ff., records earthquakes at Antioch in 115, 458, 525, 528, 557, 579, 587, to go no farther.

[52] *Constantinople*] Presumably the earthquake of 14 September 1509 (Laurentius Surius, *Commentarius brevis rerum in orbe gestarum, ab anno salutis M. D.,* Coloniac, 1586). Another earthquake rocked the city on 10 May 1556 (Mallet, *op. cit.,* p. 10).

whole yeare together,[53] that the Emperour, and all the people, were faine to dwell abroade in the fieldes, vnder tentes and pauilions for feare their houses would fall on their heades, it is recorded in Chronicles, and worthy to be remembred.

The seconde kynde is, when the earth with great violence is lifted vp, so that the buyldinges are lyke to falle, and by and by synketh downe agayne: this is when all the force of the wyndes stryueth to get vpwarde, after the nature of gonpouder, and fyndyng some waye to be delyuered out of bondage, the earth that was hoysed vp,[54] returneth to his old place.

The third kynde is a gapinge, rendyng, or cleauing of the earth, when the earth synketh downe, and swalloweth vp cities, and townes, with castels, and towers, hylles and rockes, ryuers, and floodes,[55] so that they be neuer seen again. Yea the Sea in some places hath been [C5r] dronke vp, so that men myght haue gone ouer on foote, vntyll the tyme of tyde or flood returning, couered the place with waters againe. But in the lande, where this earthquake swalloweth vp any cytie, or countrie, there apeareth nothing in the place thereof, but a marueylous wyde and deape goulf, or hole. *Aristotle* maketh mention[56] of diuers places, and regions that were ouerthrouwen with this kynde of earthquake.

The fourth kynde, is when greate mountaynes ar cast vp out of the earth, or els when some part of the lande synketh downe, and in steade thereof aryse ryuers, lakes, or fyers, breakyng out with smoke and ashes. It causeth also ouerflowyngs of the sea, when the sea bottom, is lyfted vp, and

---

[53] together] without interruption.

[54] hoysed vp] lifted up.

[55] floodes] bodies of water.

[56] mention] Perhaps Fulke refers to the *Meteorologica* 366a.

by this meanes, arise many Ilandes in the sea, that neuer were seen before. These and other suche miracles, are often to be founde in the wryters of histories, also in the *Philosophers*, as *Aristotle, Seneca*, and *Plinius*.[57]

Neuerthles, the effectes of some, as moste notable it shall not be vnprofitable to recite. *Plato* in his *Dialogue*, intituled *Timeus*, maketh mention by the way [C5v] of a wonderfull earthquake, wherebye not only *Africa* was rent asonder from *Europa* and *Asia* (as it is indead at this daye, except a lytle necke by the redde Sea,) the Sea entring betwene them that nowe is called *Mare mediterraneum:* But also a wonderfull great Ilande, whiche he affirmeth, was greater then *Aphrica* and *Asia* both, called *Atlantis*,[58] was swallowed vp, and couered by the waters, in so muche, that on the Sea called *Atlanticum;* for a great whyle after, no shippe could sayle, by reason that the same huge sea, by resolution[59] of the earth of that myghty Iland, was al turned into mudde. The famous Ile of *Scicilia* was also some tyme a part of *Italy*, and by earthquake rent asonder from it. *Seneca* maketh mention of two Ilandes, *Theron* and *Therea*,[60] that in his tyme, first apeared. It should seeme both by *Aristotle*, and also by *Herodotus*, that *Egypt*, in auncient tyme, was a goulphe of the sea, and by earthquake made a drye lande. During the raygne of *Tyberius* the Emperour, twelue notable cyties of Asia, were ouerthrowen in one nyght, etc.

---

[57] *Plinius*] Chiefly in the *Historia naturalis* ii. 89.

[58] *Atlantis*] Fulke's account of Atlantis and other lands affected by earthquakes owes much to the *Quaestiones naturales* vi. 21, 29 and to the *Historia naturalis* ii. 87-92. The attribution of the Nile delta to earthquake rather than alluvial action seems to be a misreading of the *Meteorologica* 352a and the *Historia naturalis* ii. 87.

[59] resolution] dissolution, liquefaction.

[60] *Theron* and *Therea*] better known as Therasia and Thera, of the Cyclades group in the Aegean Sea (vi. 21).

[C6r] ¶ *How so great wyndes come to be vnder the earth.*

THE great caues and dennes of the earth, must neades be full of ayere continually, but when by the heate of the sonne, the moysture of the earthe is resolued,[61] many *Exhalations* ar generate as well within the earth, as without, and where as the places were full before, so that they coulde receyue no more exepte part of that which was in them nor lett out, in suche countries, where the earth hath fewe pores, or els where they bee stopped, with moysture, it must neades followe, that these exhalations striuing to get out, must neades rende the earth in some place, or lifte it vp, so that either thei may haue free passage, or els rowm inough to abide in.

Of the signes and tokens that goe before an earthquake most commonly.[62]

THE first is the raging of the sea, when there are no tempestuous wyndes, to styre it, yea when the ayre is moste calme without wyndes. The cause why the Sea then rageth, is that the wynde beginneth to labour for passage, [C6v] that waye, and fynding none, is sent back, and soone after shaketh the lande. The seconde sygne is calmenes of the ayre, and colde, whiche cometh to passe by reason that the *Exhalation,* that shold be abroade, is within the earth.

The thirde signe, is sayde to be, a longe thine strake[63] of a cloude seen, when the skye is cleare, after the setting of the sonne. This (saye they) is caused, by reason that the *Exhalation* or vapor, whiche is the matter of cloudes, is gone into the earth Other affirme that it is the *Exhalation*

[61] resolued] dissipated.

[62] commonly] Most but not quite all of this chapter echoes the *Historia naturalis* ii. 81.

[63] strake] beam of light.

that breaketh out of som narrowe hole of the earth, out of whiche the rest of the wynde cannot issue, neither will it wayghte the tyme, wherfore within a whyle after, it seeketh and maketh it selfe by soden eruption a broader waye to be deliuered out of pryson.

Also the sunne certaine dayes before it, appeareth dimme, because the wynd, that should haue purged and dissolued the grosse ayere, that causeth this dymnes, to our eyes, is enclosed within the bowels of the earth.

The water in the botome of deape welles, is troubled, and the sauor therof [C7r] infected, because the pestilent *Exhalations* that haue ben long inclosed, within the earth do then beginne a litle to be sent abrode. For thereof cometh it, that in many places where earthquakes haue been, great aboundaunce of smoke, flame, and ashes, is cast out, when the aboundaunce of brymstone that is vnder the grounde, through violent motion is set on fyre, and breaketh forth. Finally, who knoweth not, what stynking mynerals and other poysonous stuffe doth growe vnder the earth? wherfor it is no wonder if well water, before an earthquake, be infected, but rather it is to be marueiled, if after an earthquake, there followe not a greuous pestilence, when the whole masse of infection is blowne abroade.

Last of all, there is harde before it, in the tyme of it, and after it, a great noyse and sounde vnder the earth, a terrible groanyng, and a verye thondryng, yea somtymes when there followeth no earthquake at all, when as the wynde without shaking of the earth, fyndeth a waye to passe out at. And these for the moste part, or at lest some of them, are forewarninges that the moste fearfull [C7v] earthquake wil followe, then the which there is no natural thing, that bryngeth men into a greater feare. *Cato* was very curius to confesse him self, that he repented, that euer he went by water, where as he might haue gone by lande.[64] But what lande, can be sure? if it be the Lordes will, by this woorke of his to shake it? what building so strong that can defende vs? when the more stronge the more danger, the higher the greater fall.

¶ *Of thonder.*

THonder is a sound, caused in the cloudes, by the breaking out of a whote and dry *Exhalation,* beating against the edges, of the cloude. It is often herde in spryng and sommer, by reason that the heat of the sonne, then draweth vp many *Exhalations,* which meating in the midle region of the ayre, with colde and moist *vapors,* ar together with them, inclosed in an hollowe cloud, but when the whot *Exhalation* cannot agree with the coldnes of the place, by this strife being driuen together, made stronger and kendled, it wil neades breake out which soden and violent eruption, causeth the noyse which we cal thonder. A similitude is put by gret autors of moist wood,[65] that cracketh [C8r] in the fire, we may adde here vnto the breaking of an egge in the fire, of an apple, or any like thing, for whatsoeuer holdeth and withholdeth, enclosed any whot wind, so that it can haue no vente, it wil seeke it self a way, by breaking the skinn, shell or case. It wer no ill comparison to liken thonder to the sound of a gonne, which be both caused of the same or very like causes.

The sound of thonders is diuerse, after which, men haue diuided the thonders into diuerse kindes. Making first ij. sortes, that is, small thonder and great. But as for the

---

[64] *Cato*] According to Plutarch (*Life of Marcus Cato* ix. 6, in North's translation), Cato said "in all his life time he repented him of three things. The first was, if that he euer told secret to any woman: the second, that euer he went by water, when he might haue gone by land: the third, that he had bene idle a whole day, and had done nothing."

[65] moist wood] *Cf. Meteorologica* 369a, *Quaestiones naturales* ii. 12.

diuersitie of soundes, generally it commeth of the diuerse disposition of the cloudes, one while hauing more holes then at another, somtime thicker in one place then in another. The smal or litle thonder is, when the *exhalation* is driuen from side to side, of the cloude, making a noise, and ether for the smal quantitie, and lesse forciblenes,[66] or els for the thicknes of the cloudes walles, is not able to break them, but rombleth vp and down within the cloud, whose sids ar stronger then the force of the *exhalation* is able to breake, it ronneth vp and down within, and striking against the cloud and moist sides, maketh a noyse not vnlike to the quenching of whote yron in cold water.

[C8v] And if the *Exhalation* be meanly strong, and the cloude not in all places of lyke thickenes, it breaketh out at those thinn places with[67] suche a bussing, as wynd maketh blowyng out of narrow holes.

But if the cloude, be so thynne, that it cannot kepe in the *Exhalation,* although it be not kyndled, then it bloweth out with lyke puffinge as wynde commeth out of a payre of bellowes.

A great thonder, is when the *Exhalation* is muche in quantitie, and verye whote and drye in qualitie, the cloude also very thycke and stronge, that easely wyll not geue place to the wynde, to escape out.

Wherfore if the *Exhalation* do vehemently shake the cloude, though it doe not at the first disperse it, it maketh a longe and fearefull romblyng against the sydes of the cloude, vntill at the last being made stronger by swyfter motion, it dissolueth the cloude, and hath lybertie to passe out into the open ayer. The cloude resolued,[68] droppeth downe, and then followeth a showre of rayne.

---

[66] forciblenes] 1602; 1563 forcibles.
[67] with] 1602; 1563 whiche.
[68] resolued] dissolved, disintegrated.

Other whyles it shaketh the cloude, not long, but streyghtwaye rendeth it a [D1r] long space and tyme, whose sounde is like the rendyng of a broade clothe, which noyce continueth a prety whyle.

And sometime it discusseth[69] the cloude at once, makyng a vehement and terryble cracke lyke a gonne, sometime with great force, casting out stones, but most commenly fyre, whiche setteth manye highe places on fyre. As in the yeare of our Lorde, 1561. the fourth day of *Iune,* the steple of saint Paules church in London was set on fyre,[70] as it hath been once or twyse before, and burned. The noyce of thonder though it be great in suche places ouer whiche it is made, yet is it not harde farre of, especially against the wynde. Whereof we had experience also in the yeare of our Lorde, 1561. on saynt Mathyes days in *February,*[71] at the euening, when there was a great flashe of lyghtnynges, and a verye terryble crack of thonder followynge, they that were but xv. myles from London Westwarde, hearde no noyse, nor sound therof: the wynde that tyme was Western.

The effect of thonder is profitable to men, bothe for that the swete shower doth followe it, and also for that it pur- [D1v] geth, and purifieth the ayre by the swyft mouynge of the *Exhalation,* that breaketh foorthe, as also by the sounde which deuidynge and pearcyng the ayre, causeth it to be muche thynner, which may be veryfied by an historie that *Plutarchus* in the life of *T. Quincius Flaminius,* reporteth, that there was suche a noyce made by the *Grecians*

---

[69] discusseth] dispels, disperses, scatters.

[70] set on fyre] There are several contemporary accounts of the firing of St. Paul's steeple by lightning on 4 June 1561. *See Documents Illustrating the History of St. Paul's Cathedral,* ed. W. Sparrow Simpson (London, 1880, Camden Society n.s. **26**); *A Survey of London by John Stow,* ed. C. L. Kingsford (Oxford, 1908), **1**: pp 331-332; **2**: pp. 346-347.

[71] *February*] 24 February 1562. See *Three fifteenth-century Chronicles, with Historical Memoranda by John Stowe,* ed. James Gairdner (London, 1880, Camden Society, n.s., **28**), p. 115.

after theyr lybertie was restored, that the byrdes of the ayer that flewe ouer them were seen to fall downe,[72] by reason that the ayer deuided by theyr crye, was made so thinn, that there was no firmitie,[73] or strengthe in it to beare them vp. And let this suffice for thonder, whome lyghtnyng succeadeth in treatie,[74] that seildome is from it in nature.

¶ *Of Lyghtninge.*

Among the diuerse kindes of lightning, whiche wryters in this knowledge doe nomber, we shall entreate onely of fowre kyndes, yet so, that vnder these fowre, all the reste maye be comprehended. The names we must borrowe of the latin tongue. The first is *Fulgetrum,* The seconde *Coruscatio,* the thyrd *Fulgur* the fourth, *Fulmen.*[75]

[D2r] ¶ *Of Fulgetrum.*

F*vlgetrum* we cal that kynde of lyghtening which is seen on sommer nights and eueninges, after a whote daie. The generation hereof is suche, when many thynn, light, and whote *Exhalations,* by the immoderate heate, haue ben drawen vp from the earth, and by the absence of the sunne, be destitute of that force, wherby they should haue been drawen further vpwarde, yet somethinge ascending by their owne nature, in that they be lyght and whote they meat

---

[72] Plutarch (*Life of Flamininus* x. 6) gives considerable space to the incident Fulke mentions. After his victory over Philip of Macedon at Cynoscephalae in 197 B.C., Flamininus had a herald announce at the Isthmian games that the Greek states were to be free and independent. The shouts that followed were so loud that crows flying overhead fell into the theater. Plutarch provides three possible explanations: (1) that the parts of the air separated, leaving the birds no support, (2) that the force of the sound struck the birds like an arrow, (3) that the sound produced a circular motion in the air like that of a whirlpool in the sea.

[73] firmitie] firmness, stability.

[74] in treatie] in (this) discussion.

[75] *Fulmen*] Except for *fulgere* (*Quaestiones naturales* ii. 56), these terms do not appear in Fulke's chief sources. Isidore of Seville, however, uses *fulgur* and *fulmen* (*Etymologiae* XIII. ix).

with the colde, either of the night in the lowest region, or els of the ayre in the midle region, and so by resistence of contraries (as it hath been oft before rehersed) they ar beaten back, and with the vehement mouing set on fire. This lightning commonly goeth out in the ayre, terrible to beholde, not hurtful to any thing. Except somtim when the matter of it is, earthy and grosse, being striken downe to the earth, it blasteth corne. and grasse, with other small hurt. Sometyme it setteth a barne or thacked[76] house on fyere. The collour of this lightninge, as of all other, is dyuerse, [D2^{v}] partly according to the matter, and partly accordyng to the lyght. If the matter be thynne, it is whyte, if the substaunce be grosse, it is ruddy, lyke flames of fire, in great light as in the daie it appeareth whyte, in the nyght, ruddy, yet somtime in the daye tyme, we may see it yealow, whiche is a token that the matter is wonderfull thicke and grosse. Olde wyues are, wont to saye that no nyghte in the yeare, except one passeth without lyghtnynge, but that is as true as the rest of theyr tales, whereof they haue greate stoore.

¶ *Of Coruscation.*

Coruscation is a glistering of fyre, rather than fyre in deade, and a glymmerynge of lyghtning, rather then lightning it self, which is ij. maner of wayes, one waye, when cloudes that be lower then the vpper part of the earth, without the compasse of our syght, are enflamed, and the reflexion of that flame, is cast vp into our syght appearyng in all poyntes lyke lyghtning, sauinge that the ayer where it appeareth is so cleare, that we are perswaded no lightning can be ther caused. Another waye, is when there be thycke cloudes ouer vs, and commonlye [D3^{r}] a double order of

[76] thacked] thatched.

cloudes, one aboue another, if lightninge or any other inflammation be in the vpper part of these clouds, the lyght of them perceth through the lower partes, as through a glasse, and so appeareth as though it lyghtned, when perhappes it did lyghten in deade, yet that whiche we sawe, was but the shadowe therof. And this is often without thonder.

¶ *Of Fulgur.*

FVlgur is that kind of lightning which followeth thonder, whereof we haue spoken before. For when that violent *Exhalation* breaketh foorthe, makynge a noyce as it beateth against the sydes of the cloude, with the same violence, it is set on fyre, and casteth a great lyghte, whiche is seen, farre and neare. And although the lyghtnyng appeare vnto vs, a good preaty whyle before the thonderclappe be harde, yet is it not caused before the noyce if any thonder at all doe followe, but eyther is after it or with it.[77] Wherfore that[78] we see it, before we hear the thonder, may be ascribed, either to the quicknes of our syght, that preuenteth[79] the hearing, or els to the swyft mouinge of [D3v] the fyer and the lyght thereof, to oure eyes, and the slow mouynge of the sound vnto our eares and hearynge. These three kyndes of lyghtnyges, are more feareful then hurteful, but the fourth seldome passeth without som damage doing.

¶ *Of the fourth kynde called Fulmen.*

THe moste dangerus, violent, and hurtfull, kinde of lightning is called *Fulmen,* whose generation is suche, as followeth. What tyme, a whote *Exhalation,* is enclosed in a cloude, and breakynge the same, bursteth foorth, it is set

---

[77] or with it] So said the *Meteorologica* 369b and the *Quaestiones naturales* ii. 12. Pliny (ii. 55) held that the two phenomena occurred simultaneously.

[78] Wherfore that] the reason why.

[79] preuenteth] acts before, precedes.

on fyre, and with wonderfull greate force stryken downe toward the earthe: The cracke of thonder, that is made when this lyghtenynge breaketh out, is sodayne, shorte, and greate, lyke the sounde of a gonne. And often tymes a greate stoone is blowne out, with it, which they call the tonder bolt, which is made on this manner. In the *Exhalation,* whiche is gathered out of the earthe, is muche earthy matter, with clotterynge[80] together by moysture, beyng clammy by nature, consistynge of brymstone and other metallycke sub- [D4r] stance, by the excessiue heate, is hardened as a brycke is in the fyere, and with the myghtye force of the *Exhalation,* stronglye cast towarde the earthe, and stryketh downe steples, and hyghe buildynges of stoone, and of woodde, passeth through them and setteth them on fire, it cleueth trees and setteth them on fire, and the stronger the thynge be that resisteth it, the more harme it dothe to it.[81] It is sharpe poynted at one ende, and thycke at the other ende, whiche is caused by reason, that the moyster part, as heauyer, goeth to the bottome of it. So is the toppe smal, and the bottom thick.

Men wryte that the thonderbolt goeth neuer aboue fiue foote deepe, when it falleth vpon the earth, whiche standeth with reason, both because the strengthe of it is weakened, before it com so neare the ground, and also, because the continual thicknes of the earth, breaketh the force, were it neuer so great.

Both *Aristotle, Seneca* and *Plinius,* deuide this lyghtning into three kyndes.[82]

❡ *Of the fyrst.*

THe first is drye, whiche burneth not, to be felte, but deuideth, and perceth [D4v] with wonderfull swyftnes.

[80] clotterynge] coagulating.

[81] dothe to it] *Quaestiones naturales* ii. 52.

[82] kyndes] Fulke is actually following Pliny (ii. 52). Seneca ii. 40 is relevant but not identical, and the *Meteorologica* makes no clear division.

For beinge subtyle and pure, it passeth through the pores of anye thynge, be they neuer so small, and such thynges, as giue place vnto it, it hurteth not, but suche thyngs as resisteth, it deuideth and perseth. For it wyll melt mony in mens purses, the purses being whole, and vnharmed.[83] Yea, it wyl melte a swerde in the scabberde, and not hurt the scabberd at al. A wyne vessell it wyll cleaue, and yet the wyne shalbe so dull,[84] that by the space of three dayes it wyll not ronne out. It wil hurt a mans hande and not his gloue. It wil burne a mans bones within hym to ashes, and yet his skynne and fleshe shall appeare fayre, as though nothinge had commen to hym. Yea otherwhyle the whole man in the moment of an howre, shalbe burned to ashes, where as his clothes, shal not seeme to haue been touched. It wyll also kyll the chylde in the mothers belly, and not hurte the mother. And all because the mater is very subtyle, and thinne, burnyng, and passinge through whatsoeuer it be, that wyll not geue it free passage.

ℭ *Of the second kinde.* [D5r]

THE seconde kynde is moyste, and because it is very thinne, it burneth not to ashes, but only blasteth, or scorcheth trees, corne, and grasse: and by reason of the moistnes, it maketh all thynges black, that it commeth neare, as moyste wood burning, is smokye and maketh thynges neare it to be blacke and smokie.

ℭ *Of the thirde kinde.*

THE thirde kynde is moste lyke oure commen fyre, that we haue here on the earth of grosse and earthly substance, wherfore it leaueth a prynte where it hath been, or els consumeth it into ashes, if it be suche a body as wylbe burned with fyre.

---

[83] vnharmed] *Quaestiones naturales* ii. 31-32; *Historia naturalis* ii. 52.
[84] dull] inert, sluggish.

¶ *Of the maruayls of Lightening and their causes.*

BEside the wonderful effects of lightnyng, that haue been already remembred, there be many other whiche hereafter ensue, with the reason and causes vnto them belonging, as thus.

The nature of lyghtning is, to poyson beastes that are stryken therewith, as though they had been bytten by a serpent. The cause of this is, that the matter of lightnyng, is muche infected with [D5v] brymstone, and other poysonous metallike substances, whiche will poyson the rather in lightening, because it is thinn and geueth them passage into euery part of the body. It is notable, that *Seneca* wryteth,[85] howe wyne vessels of wood beinge burned with lightning, the wyne wold stande styll, and not runne out, the reason hereof, is the swyfte alteration and chaunge, wherby also, all the clammynes of the wyne, is drawen to the outward moste part, and so keepeth in the wyne, as in a skynne, that by the space of three dayes, it wyll not ronne. It wyl also poyson wyne, in so muche that they whiche drynke thereof, shall eyther be madde or dye of it. The cause hereof was set foorth before.

Lightning that striketh a poysenous beaste purgeth it from the poyson, in so muche that it causeth a serpent or snake whiche it kylleth, to breade wormes, whiche otherwyse it would not doe, but beyng purged from the naturall poyson by the swyfte percyng of the lightning: nothyng letteth.[86] but that it may breade wormes,[87] as all other corrupte fleshe wyll doe.

[D6r] If lyghtning strike one that slepeth, it openeth his eyes, and of one that waketh, it shytteth the eyes.[88] The

[85] *Seneca*] *Quaestiones naturales* ii. 31, 53.

[86] letteth] prevents, stands in the way.

[87] wormes] *Quaestiones naturales* ii. 31.

[88] shytteth the eyes] *Historia naturalis* ii. 55. *Shytteth* is an alternate form of the present tense of *shut*, from O.E. *scyttan*.

cause is this, that it waketh hym that sleepeth, and kylleth hym before he can close his eyes agayne. And hym that waketh, it so amaseth, that he wynketh, as he wyll doe at any sodayne chaunse, so he dyeth before he can open his eies agayne.

All lyuynge thynges, turne their face towarde the stroke of the lyghtenyng, because it is their nature, to turne their head if any thing com sodenly behind them. The reste that haue theyr face toward it, when it commeth, neuer turne before they be kylled.

The reason why it kylleth the child in the mothers wombe, not hurtynge the mother, is the tendernes of the one, and the strengthe of the other, when the lightenyng is not vehement, otherwyse both should dye together.

Sometyme lyghtening burneth only the garmentes, shoes, or heare of men, not hurtynge theyr bodyes, and then the *Exhalation* is nothyng vehement. Some time it kylleth a man and there apereth no [D6v] wounde without, neyther anye hurte within, no not so muche as any signe of burnyng: for then the *Exhalation* whiche being kindled is called lyghtning is wonderfull subtill and thinne, so swiftly passing through that it leaueth no marke or token behinde it.

They that beholde the lightening, are either made blynd, or their face swelleth, or els become lepers, for that fyery *Exhalation,* receiued into the pores of their face and eyes, maketh their face to swel, and breake out into a leprosie, and also drieth vp the Christalline humor of their eyes, so that consequently they must needes be blynde.

Eutropius sheweth,[89] that the same day in whiche *Marcus Tullius Cicero,* was borne, a certeine virgine of *Rome* ryding into *Apulia,* was striken with lightening, so that all her

---

[89] *Eutropius* sheweth] It is not in the *Breviarium* of Eutropius but in the *Historia Romana* of Paulus Diaconus, which was based upon it and sometimes printed with it, that this story is found (*Pauli Diaconi Historia Romana,* ed. Amedeo Crivellucci, Roma, 1914, Fonti per la storia d'Italia [no. 51], p. 70).

garmentes beinge taken from her without any rendinge, she laye starke naked, the lasing of her brest being vndone, and her hose garters vntied: yea, her bracelettes collers and rynges, being also loosed from her. Lykewyse her horse laye dead with his bridle and girtes[90] vntied.

[D7r] The places of them that are burnt with lightning are colder then the reste of their bodies, other[91] because the greater heat draweth away the lesser, or els because, that by the great violence the vitall heate is quyghte extinguished in that place.

The sea Calfe is neuer hurt with lyghtening, wherfore the Emperoures tentes, were wonte to be couered with their skinnes.

The Baye trees, and boxe trees, are neuer or seldom stryken with lyghtning. The cause of these may be, the hardnes of their skinne, which hath so fewe pore holes, that the *Exhalation* can not enter into them.

The eagle also among fowles is not stryken with lightening, wherfore the *Poetes* fayne, that the Eagle carieth *Iupiters* armur, whiche is lightnyng. The reason may be the thicknes and drienes of her fethers, whiche wyll not be kindled with so swift a fyre.[92]

¶ *Of storme wyndes.*

A storme wynde, is a thycke *Exhalation* violently moued out of a cloude without inflammation or burning. The mat-[D7v]ter of this storme, is all one with the matter of lightening, that hath been spoken of: namely it is an *Exhalation* very whot and drye, and also grosse and thycke, so that it wyll easely be set on fyre, but then it hath another name, and other effectes.

---

[90] girtes] saddle-girths.

[91] other] either.

[92] swift a fyre] With this and the two preceding paragraphs *cf. Historia naturalis* ii. 56.

The forme or maner of the generation is suche. When abondance of that kynde of *Exhalation* is gathered together, within a cloude, whiche nedes wyl haue one waye out or other: it breaketh the cloude, and causeth thonder, as it hath been tought before, but if the matter be very thicke, and the cloude somewhat thynne, then doth it not rende the cloude, but fallynge downe, beareth the cloude before it, and so is caried as an arrowe out of a bowe. It doth alwayes goe before a great soden showere, for when the cloude is broken, the water muste needes fall downe. Also it is so grosse, and so thicke, that it darkeneth the ayre, and maketh all the lowest region of the ayre, to be in manner as a darke smokye cloude. It causeth tempeste in the Sea, and wonderfull great daunger to them that beare sayle, whome if it ouer-[D8r]take, it bryngeth to vtter destruction. So soden it is, that it can not be resisted wyth sodeyn helpe. So violent it is, that feble force canne not withstande it. Finally, it is so troublesome wyth thonder, lyghtnynge, rayne and blaste, besydes these darkenesse and colde, that it woulde make menne, at so neare a pynche to bee at their wyttes endes, yf they weare not accustomed to suche tumultuous tempest. Wherfore it weare profitable, to declare the signes that go before it, to the ende, menne myght beware of it. But they are so commen to other tempestes, that either they are knowen well enoughe, or els beynge neuer so well knowen, in a seldome[93] calamytie they woulde lytle bee feared. The Sea shyppes subiecte to more danger, haue more helpe if it bee vsed in tyme, but no sygnes foreknowen, can profit the dweller of the lande, to keepe his house from ruine, except it weare to saue his lyfe from the fall of his mansion. The soden violence of this tempest to hym, is more seldom tymes, but more incurable when it commeth

[93] seldome] infrequent.

then to the Marryner, who hathe some ayde [D8^{v}] to looke for, by his comming, the other if he escape with his lyfe, may comfort hym selfe, that he was neare a greate daunger, and cast with hym selfe[94] to builde vp his house agayne.

¶ *Of whyrle wyndes.*

A Whirlewinde, is a wynde breaking out of a cloude. rowling or wynding round about, ouerthrowyng that which standeth neare it, and that whiche commeth befor it, carying it with him a loft in the ayre.

It differeth from a storme wynde in thre pointes.

First in the matter which is lesse in quantitie, and of thinner substaunce.

Secondly in the mouing, whiche is circulare wyndyng about, where as the storme bloweth a slope and sydelonges. Also a whyrlewynde in the mouinge diuideth not it selfe abroade, and bloweth dyrectly as the storme doth.

And thirdly in the maner of the generation, for a storme doth alwayes come out of one cloude, but a whyrlewynde some tyme is caused by meanes of twoo contrary wyndes that meete together. In lyke maner, as we see in the streates [E1^{r}] of cyties, where the wynde is beaten back from two walles, meetinge in the myddest of the streate, there is made a lytle whyrlewynde, which whiskynge round about taketh vp the dust, or strawes and bloweth it about after the very similitude of the great and feareful whirle wynde. The reason of the going about, is this, that when the walles beat back the wynde from them, whiche aboundeth in that place, and those wyndes, when they meete by reason of equal force on bothe sydes, can neyther dryue one the other back agayn, nor yet passe through one the other: it must needes be, that they must bothe seeke a waye on the syde at once, and consequently, be caried round about, the one as it were pur-

[94] cast with hym selfe] made plans (*New English Dictionary,* Cast *v.* 42).

suing the other, vntyll there be space enough in the ayre, that they may be parted asonder.

The matter of a whyrlewynde, is not muche differing from the matter of storme and lyghtening, that is an *Exhalation* whote and drye, breakyng out of a cloude, in diuerse partes of it, which causeth the blowyng about, also it is caused as it hath been sayde, by twoo or more wyndes, blowyng from diuerse places, [E1v] whiche may be of particular causes that hath been shewed[95] before in the chapter of wyndes. This tempest is noysome[96] to man and beaste, Sea and lande, thyngs lyuing, and life lacking. For it wyl take vp bothe men and beastes, stoones and cloddes of earth, whiche when it hath borne a great waye wyll not be so curteus[97] as to sette them downe agayne, but neglygently letteth them fall from a great heyght, or els violently throweth them downe to the earth.

It breaketh trees wyndyng them about and pulling them vp by the rootes. It turneth about a shippe and brooseth[98] it in peaces with other mischiefes beside.

¶ *Of the fyred whirle wynde.*

Sometyme a whirlewynde, is sett on fyre within the cloude, and then breakyng foorthe, flyeth rounde lyke a great cartewhyle, terrible to beholde, burnynge and ouerthrowynge all drye thinges, that it commeth neare, as houses, woodes, corne, grasse, and whatsoeuer els standeth in the waye.

It differeth not from a whirlewind, sauing that it is kindled and set on fyre, so apearing, els the generation of both is [E2r] called one.

[95] shewed] 1602; 1563 sheweth.
[96] noysome] noxious, harmful.
[97] curteus] 1602 courteous; 1563 curtues.
[98] brooseth] crushes.

¶ *Of Circles.*

THE Circle called *Halon,*[99] is a garland of diuerse collours that is seen about the sunne, the Moone, or any other sterre specially about *Iupiter* or *Venus,* for their greate bryghtnes. It is called of the Greeks a compassed[100] platte, of the Latines a crowne or garlande.

The matter wherin it is made,[101] is a cloude of equall thicknes, or thinnes, comming directly vnder the body of the sunn, the Moone, or other sterres, into whiche the lyght of the heauenly body is receyued, and so appeareth rounde, because the sterre is rounde, or as a stoone caste into the water, maketh many round circles, dilatyng in breadth, vntyll the violence of the mouyng is ended: so is it in the ayre the lyght beames percynge it, causeth broade Circles to be delated,[102] whiche appeare whyght, purple, black, redde, greene, blewe, and other collors, according to the disposition of the cloudes mater. The cause of suche collours, is shewed before in the peculiar treatie of collours.[103]

This circle is oftener seen about the Moone, then about the Sunne, because [E2v] the heate of the Sunne draweth the *vapors* to hyghe, where it can not be made. Also, because the nyght is a more quiet tyme then the daye from wynde, it is more often in the nyght, then in the daye. Syldome about other sterres, because their lyght beams ar to weake often to perse a cloud, yet oftner about smal sters then the Sunne, because the lyght of the Sunne, perceth the cloude more forcibely, then that this *Halon* can many tymes be cause.[104]

Otherwhyles it is seen about a candell, which must be in a very thicke and grosse ayre, of suche proportionate thicknes,

---

[99] *Halon*] halo.
[100] compassed] curved.
[101] The matter wherin] *Cf. Meteorologica* 372b, *Quaestiones naturales* i. 2.
[102] delated] dilated, spread out.
[103] treatie] treatment.
[104] then . . . cause] so that this halo cannot many times be cause.

that it may receiue the lyght as the cloude doth from the sterres, as in smoky places, or whotehouses.

This kynde of Circle, is sometimes lyke a raynbowe, sauynge that it is a whole circle, vnlesse the sterre vnder whiche it is caused, be not all rysen, or els the cloude in whiche it is seen be not al come vnder the sterre, or after it hath come vnder some parte thereof be dissolued[105] from the rest.

These Circles be signes of tempest, and wyndes, as wytnesse bothe *Virgile,* [E3r] and *Aratus.*[106]

The wynde shall blowe from that quarter, where the circle first beginneth to breake. The cause whereof is this, that the circle is broken, by the wynde that is aboue, whiche is not yet come downe towardes vs, but by this effect aboue, we may gather both that it wyll come, and also from what quarter.

A great Circle about the Moone, betokeneth great colde and frost to follow after.

But if it vanyshe awaye and be dissolued altogether, it is a signe of fayre weather.

If it be brooken in many partes, it signifieth tempest.

If it wax altogether thicker, and darker it is a fore warnyng of rayne.

One alone after *Ptolomee,* pure and whyte, vanyshing away by lytle and litle, is a token of fayre weather.

Twoo or three at once, portendeth tempest, if they be ruddy, they shewe wynde to come, and toward[107] snowe, they seeme as it were broken and rockye.

Being darke or dymme, they signifie all these forsayde euentes, with more [E3v] force and abundaunce, it is

---

[105] dissolued] detached, released.

[106] *Aratus*] Vergil's account of the weather signs appears in the first book of the *Georgics* (as Fulke says on p. 93). The reference to Aratus (3d century B.C.) the author of a poem on the heavens, could have been derived from the *Quaestiones naturales* i. 13.

[107] toward] in prospect of, in the imminence of.

oftener caused in *Autumne,* and spring then in wynter or sommer, the cause is the temperatnes of the tyme.

The cause why it apeareth somtime greater, and sometyme lesser, is in the qualitie of the matter, whiche as it is grosse, or thynne, wyll more or lesse be dylated, and stretched abroade, and also as some wil haue it, of the weakenes of mens syght. Of whiche *Aristotle* bryngeth an example in one *Antipho,*[108] whiche dyd alwayes see his owne image before hym in the ayre, as in a glasse, whiche he affyrmeth to haue been for the weakenes of his syght beames, that coulde not pearce the ayre, so that they weare reflected agayne to hym selfe.

And thus much for *Halone* and the causes, signes, or tokens of it.

¶ *Of the Raynbowe.*

THE Raynbowe, is the aparition of certain collours in a cloude opposite against the sunne, in fashion of halfe a Circle. *Possidonius*[109] sayde, it was the sunnes lookyng glasse, wherein his image was represented, and that the blewe [E4r] colloured, was the proper collour of the cloude, the redde of the sunne, all the other collours of commixtion.

It differeth manifoldly from *Halone,* for the raynbowe is alwayes opposite against the sunne, but *Halone* is directly vnder it.

They differ not onely in place, but also in fashion, the raynbowe, is but halfe a Circle, the *Halon* is a whole Cyrcle.

Lykewyse they vary in colloure, for the raynebowe is more dymme and of purple collour, the *Halone* whyter and bryghter.

Also in continuance, for the raynebowe may continue, longer, then *Halone.*

---

108 *Antipho*] *Cf. Meteorologica* 373b, although Antiphio is not named there.

109 *Possidonius*] This may come from the *Quaestiones naturales* i. 5. The writings of Posidonius are not extant.

The image of the raynbowe may be seen on a walle, the sunn striking through a sixe pointed stoone, called *Iris,*[110] or anye other Christall of the same fashion, also through some glasse wyndowe.

Halone is seen aboute candelles, in smoky places, as are bathes and kychenes.

The manner of the generation of the raynbowe is suche, there is opposit againste the sunne, a thycke watrye cloude, whiche is alreadye resolued into [E4v] dewye droppes of rayne, as (for a grosse similitude) is seen on the potlidde when the water in the vessell hath sodden,[111] or is very whote, the lydde wylbe al full of small droppes of water, whiche come from the water in the vessell, fyrst by heat resolued into smoke, after when it can not goe at large, it is resolued agayne. Wherfore vpon such a cloude, the sunne beames strykynge, as vppon a smoothe glasse, doe expresse the image of the sunne vnperfectly, for the great distance. Or els the sunne beames, strike into an hollow cloude, where they are refracted or broken, and so cometh to the eyes of hym that beholdeth the raynbowe.[112]

The similitude thereof is seen, when men sayle or rowe in boates, the sunne shyneth vpon the water, whiche casteth, on the vessels syde, the collours and image of the raynbowe.

Lykewyse water in an vrinall holden against the sunne, receyueth the lyght and sheweth collours on the walle.

There be two kindes of rainbowes, one of the sunne, another of the Moone, the one by daye, the other by nyght, the raynbowe of the sunne often, but of the [E5r] Moone very seldome, in so muche that it can be but twyse at the mooste, in fiftye yeares, and that when the Moone, is in the East or West, full in perfect opposition. It hath not been

---

110 *Iris*] some kind of prismatic crystal; *cf. Historia naturalis* xxxvii. 52.
111 sodden] boiled.
112 the raynbowe] *Meteorologica* 371b, *Quaestiones naturales* i. 3-6.

many tymes seen sence the wryting of histories, yet some tymes and for the rarenes, is taken for a great wonder. Yet is it in collour nothyng so beutiful, as the sunnes, but for the moste part, whight, as mylke, other diuersities of collours are scant perceyued. When it appeareth, it is sayd to signifie tempest.

The tyme of the raynbowe, is often after the poynt[113] of *Autumne,* both for the placing of the sunne in competent lownes, and also for abundance of matters, seldome or neuer is the raynbowe seen about the middest of sommer.

There may be many raynbowes at one tyme, yet commenly but one pryncipall, of whiche the rest are but shadowes, and images, the seconde shadowe of the first, the thyrde of the seconde, as appeare by placing of their collours.

It remayneth to shew why it is but halfe a circle, or lesse, neuermore, and why the whole cloude receyueth not the [E5v] same collours, that the raynbowe hath. The cause of the fyrste is, because the center, or mydle point of the raynbowe, that is *Diametrally* apposite to the center of the same, is alwayes either in the *Horrizon* (that is the circle cutting of our sight of heauen by the earth) or vnder it. The cause why the whole cloude is not colloured, is because that in the myddest the beames as strong, perse throughe, but on the edges where they are weaker, they are reflected or refracted.

Nowe for so muche, as God made the raynbowe a signe and Sacrament of his promyse,[114] some thynke it was neuer seen before the floude. Theyr reason maye be this, that the earthe after the fyrst creation was then so fruictfull, that it neaded none, or very lytle rayne, so that suche darke cloudes, weare not often gathered, the fruictfull ground not

---

[113] poynt] (presumably) middle (*cf. New English Dictionary,* Point *sb.*[1] 18d).
[114] promyse] *Cf.* Genesis vi. 9-16.

so easely remitting[115] his moysture, that then was fatte and clammye, harde to be drawen vp: so it myght be that ther was no raynbowe before, as we cannot fynde that euer it rayned before. But whether it were or not, it is certayne, that then it became a Sacrament, wher [E6^r] as it was none before, which when we beholde, it behoueth vs to remember, the truthe of God in all his promyses, to his glorie and our comfort.

¶ *The mylke waye,*[116] *called of some the waye to saint Iames and Watlyng streate.*[117]

THe milke way, is a whyte circle seen in a cleare night, as it were in the firmament, passing by the signes of *Sagittarius* and *Gemini.*

The cause thereof, is not agreed vpon among *Philosophers,* whose opinions I thought best to reporte, before I come to the moste probable causes.

First of all *Pithagoras,*[118] is charged with a Poeticall fable, as though it had been caused by reason that the sunne did once runne out of his pathway, and burned this part wherof it loketh whyte.

Other as *Anaxagoras* and *Democritus,* sayde, that it was the light of certeine sterres, shining by them selues, of their owne light, which in the absence of the sunne, might be seen. But this opinion is also false, for the sterres haue no light of them selues, but of the sunne, also if it wer so, it shold apear about other sterrs.

---

[115] remitting] releasing, yielding up.

[116] the mylke way] the Milky Way (*New English Dictionary,* Milk *sb.* 10a).

[117] *Watling streete*] see the *New English Dictionary* (Watling-street 2), which adds: "The Milky Way received other popular names from highways, esp. pilgrimage routes. . . . A widespread designation of this kind was *the way of St. James* (of Compostella)."

[118] *Pithagoras*] The *Meteorologica* i. 8 probably gave Fulke his knowledge of Pythagoras, Anaxagoras, and Democritus.

[E6v] *Democritus* is also reported to haue sayde, that it was nothing els but innumerable lytle starres, whiche with their confuse[119] lyght, caused that whytnes, to this opinion, *Cardane*[120] semeth to subscribe.

The *Poetes* haue fowre fables of it, one that *Phaeton,* whiche on a tyme guided the Chariot of the sunn, and wandring out of the way, did burne the place, wher fore of *Iupiter* he was stryken downe wyth lyghtnyng.[121]

The second, that it is the high strete in heauen, that goeth streight to *Iupiters* pallace, and both sydes of it, the commen sorte of Gods do dwell.[122]

The thirde, that *Hebe,* one which was *Iupiters* Cupbearer, on a tyme, stombled at a starre, and shedde the wyne or mylke, that was in the cuppe, which colloured that part of heauen to this daye, wherfore she was put out[123] of her office.[124]

The fourth, that *Apollo* stoode there to fight against the Giantes, which *Iupiter* made to appeare, for a perpetuall memory.[125]

Theophrastus[126] a Philosopher affirmed, that it was the ioyning together, or seeme of the two halfe globes, whiche

---

119 confuse] confused.

120 *Cardane*] *De subtilitate* iv (*Opera omnia,* cura Caroli Sponii, Lugduni, 1663, 3: p. 417).

121 with lyghtnyng] This explanation of the origin of the Milky Way is mentioned by Aristotle (i. 8) and, among the poets, by Chaucer (*House of Fame* ii. 935 ff.).

122 Gods do dwell] Ovid, *Metamorphoses* i. 168 ff.

123 put] 1602; 1563 pout.

124 her office] S. K. Heninger (*Handbook,* p. 104) says that he has been unable to find a classical source for this fable. The only "poet" who mentions it is E.K. in his gloss to the November eclog of *The Shepheardes Calender,* which is less explicit than Fulke: "Nectar [is feigned] to be white like Creme, whereof is a proper tale of Hebe, that spilt a cup of it, and stayned the heauens, as yet appeareth." There are other references to Hebe's mishap, but they do not offer it as an explanation of the Milky Way.

125 perpetual memory] "Fulke here refers to the account of the Milky Way given by Pontano, *Meteororum liber* (in *Opera*) (Venice, 1513), fol. 132v" (Heninger, *Handbook,* p. 105).

126 *Theophrastus*] *Cf.* Macrobius, *Commentarii* I. xv. 4.

made it [E7r] appeare more light in that place then in other.

Other sayde, it was the reflexion of the shyning light of fyre, or sterre light, as it is seen in a glasse, but then it sholde be moueable.

Diodorus[127] affirmed, that it was heauenly fyre, condensede or made thick, into a circle, and so became visible, wheras the rest for the purenes, clearenes, and thinnes, could not be seene.

Possidonius[128] whose mynde to many seemeth very reasonable, saide: it is the infusion of the heate of sterres, whiche, therfore is in a circle, contrarie to the *Zodiake,* (out of whiche the sunne neuer wandreth) because it myght temper the whole compasse with vitall and lyuely heate. Although in my mynd he hath rather expressed the finall cause, then the efficient.

Aristotles opinion[129] is, that it shold be the beames of a great circle, whiche is caused by a cloude or *Exhalation* drawen vp by those sterres, whiche be called *Sporades.* This opinion of *Aristotles* is myslyked of moste men, that haue trauayled in this science and worthely. For if it [E7v] were of the nature of elementes, as *Exhalations* are, it would be at length consumed. But this circle neuer corrupteth,[130] therfor it is not of *Exhalations.* Also it nether increaseth or diminisheth, which is a playne proofe, that it consisteth not of elementall matter. Although *Aristotle* seeme to make a double circle, one celestiall, another elementall.

The last opinion is of them that say, it is of the nature of heauen, thycker in substaunce then other partes of heauen be, hauing some lykenes to the substance of the Moone, which being lightned by the same, as al the starres be,

---

[127] *Diodorus*] *Cf.* Macrobius, *Commentarii* I. xv. 5.

[128] *Possidonius*] *Cf.* Macrobius, *Commentarii* I. xv. 7.

[129] *Aristotles* opinion] *Meteorologica* 346a.

[130] corrupteth] is decomposed.

apereth whight.[131] And this opinion I take to be the most probable, because that sentence[132] of sterrelight seemeth not so reasonable to be only in that place, and not els where.

The finall cause of this milke whyte circle, hath been already touched in the opinion of *Possidonius,* wherunto also in *Plinius* in the xviij. booke and xxix. chapter of his naturall history agreeth, affirming that it is very profitable, for the generation and frutefull increase of thinges that growe on the earth. The *Mathematicians* that haue measured the breadth therof, affirme [E8r] that toward the north it passeth ouer the eclipticall lyne of the ninth spheare, from the xviij. degree of *Gemini* vnto the seconde degree of *Cancer* which is xiij. degrees, and toward the South, from the viij. degree of *Sagittarius* to the xiij. degree of the same signe, and because it is ther diuided into ij. branches (as may easly be seen in a clear night) it reacheth from the xxiiij. of *Sagittarius* to the second degree of *Capricorne.*

This circle if it be of the nature of heauen, is vnproperly placed among *Meteores* or impressions, but because of *Aristotles* mynde, who wyll haue it to be an impression kendled, and their opinion whiche thinke it proceadeth of the light of sterres it is not without good cause in this place intreated of.

¶ *Of beames or streames of light appearing through a cloude.*

THer is yet another kind of impression caused by the beames of the sunne, stryken through a watry cloud, being of vnequal thines, that is thinner in one part then in another, so that it cannot receiue the beames in any other form, then that they appere direct or slope downward, of

[131] apereth whight] Plutarch (*De placitis philosophorum*) ascribes this opinion to Parmenides.

[132] sentence] opinion.

diuers collors, and the same that ar the collors of the raynbow, though not so euident, because the re- [E8v] flexion is not so strong. They varie in collours, some ar more purple, or ruddy, when the cloud is thicker som yealow and whitish when the cloude is thinner, and so other collors ar caused likewise, wherof you may reade the proper causes in the collours of cloudes and other lyke parts of this treatise. The common people cal it the descending of the holy ghost, or our Ladies *Assumption,* because these things are painted after suche a sort. Other say that it is rayne, stryking downe in another place, as though they could see the droppes fallyng. And they are not altogether deceiued, but in the time, for sone after it wyll rayne, because this impression appeareth out of a watry cloude. They are called by dyuerse names, as roddes, wandes, coardes of tents, vnto whiche they are not much vnlyke, staues and lytle pyllers, when they seeme greater and thicker, many beyng ioyned together. The rayne bowe, the circles and these lyghtbeames, are all of one maner of generation, in so muche that if you deuide the circle, it shalbe a rayn bowe, if you drawe it streyght, in lengthe, it maketh streames or bea-[F1r]mes. Herein they agree, namely in form and matter, but they differ in outwarde forme, whiche we may call fashion, as the one is round, the other half round, and the third directe, streyght or fallynge aslope. Also they differ in place, aboute whiche they stande, for streames are only about the sunne, raynbowes about the sonne often, and seldome about the Moone, but circles both about the sunne and the Moone, and also about any other of all the sterres, yet rather and oftener about bryght sterres. To make an ende of these streames, they apeare diuersly, after the fashion and place wherein the cloude hangeth in respect of[133] the sunne. For some tyme they are seen only in the edge of a cloude, all the breadth of

[133] in respect of] in relation to.

that cloude. Sometyme through the middes of a cloude, being thynner there, then in other partes, and then they are spreade rounde about lyke a tente or pauilon vsed in warre. They ar moste commenly seen in suche tymes, as there is great aboundance of rayne, whiche they, by their apparition doe signifie not yet to be ended. And thus muche concerning direct lyght beames called roddes etc.

[F1v] ¶ *Of many Sonnes.*

IT is strange and marueilous to beholde, the lykelyhode of that, whiche *Alexander* the great,[134] sending woorde to *Darius,* sayde to be impossible, that two sonnes should rule the worlde. But oftentimes, men haue seen, as they thought in the firmament, not only two sonnes, but oftener thre sunnes, and many more in nomber, though not so often appearing. These how wonderful soeuer they appeare, proceade of a naturall cause, whiche we will endeuour to expresse. They are nothing els but Idols, or Images of the sunne, represented in an equall smooth and watry cloude, placed on the side of the sunne, and sometime on both sydes, into which the sunn beames being receiued as in a glasse, expresse the likenes of fashion and light, that is in the sunne, appearing as though there were many sunnes, where as in dede there is but one, and all the rest are images. This thicke and watry cloude, is not sayde to be vnder the sunne, for then it wolde make the circles called crownes or garlonds, it is not opposit to the sunn, for then wold it make the rainbow, but it is sayd to be [F2r] on the side, wher the image may be best represented. Also it may not be to far of, for then the beames will be to feble to be reflected, neither yet to neare, for if it so be the sunne wil

---

134 *Alexander* the great] *Cf.* Plutarch, *Apophthegmata regum et imperatorum* 180B.

disperse it: but in a competent and midle distance, for so representation of many sunnes is caused.

They are most often seen, in the morning and euening, about the rysing or going downe of the sunn, seldome at noone tyme, or about the middest of the day, because the heat will soon dissolue them. Yet hath there been some seen, which began in the morning, and continued all the daye long, vnto the euening. Sometimes ther apeare many litle sunnes, like vnto litle starres, which are caused after the same sort, as we doe see a mans face, to be expressed in all the peces of a broken glasse. So when the cloude hath many separations, there appeare many sunnes, on one syde of the true sunne, sometimes great, and sometimes litle, as the parts of the cloude separated are in quantitie.

They doe naturallye betoken tempest, and rayne, to followe because they can not appeare, but in a watry disposition of the ayre.

[F2^{v}] Also if they apeare on the Southside of the sunne they signifie a greater tempest then if they appeare on the Northsyde. The reason is alledged, because the Southerne *vapor* is sooner resolued into water, then is the Northerne.

For a supernaturall signification, they haue often tymes been noted to haue portended, the contention of Princes for kingdomes. As not longe before the contention of *Galba, Otho,* and *Vitellius,*[135] for the Empire of Rome, ther appeared three sunnes. Also of late toward the slaughter of *Lewes* kyng of *Hongary,* were seen three sonnes, betokening thre prynces that contended for the kyngdom, namely Ferdinando nowe Emperour, John Vayuode, and the great Turke.[136]

---

[135] *Galba, Otho,* and *Vitellius*] successively emperors of Rome 68-69 A.D.; Otho committed suicide, the others were assassinated.

[136] After the death in battle of King Louis (1526), the Archduke Ferdinand of Austria, John Zapolya, voivode (i.e. local ruler) of Transylvania, and Suleiman the Magnificent, sultan of the Ottomans, claimed his kingdom.

¶ *Of many Moones.*

After the treatie of many sunnes, it weare not harde for any man, without farther instruction to knowe the naturall cause of many Moones. For they are lykewyse Images of the Moone, represented in an equall cloude, which is watry, smothe, and polyshed, euen lyke a glasse. Some call them (as *Plinius* saieth[137]) night sunnes, because they ioyned with [F3r] the light of the true Moone, geue a great shynning light, to dryue awaye the shadowe and darkenes of the nyght.

It were superfluous to wryte, more of their causes, or effectes, whiche are al one with those, that haue been declared of the sunnes.

It may be doubted why the other starres doe not lykewyse expresse theyr image, in watry cloudes, and so the nomber of them to our sight should be multiplied: it may be aunswered, that their lyght or beames, are to feble, and weake to expresse any suche similitude or lykenes in the watry cloudes. For although they haue garlandes, or circles, aboute them, that are caused in a *vapor,* that is vnder them: yet it is manifest, that this apparition hath not neede of so strong a lyght, as is requyred to prynt the images of them in the cloudes. Agayne the garlandes are direct vnder, and therfore apter to receyue suche apparition.

It may be agayne obiected that the starres haue their image perfectly and sufficiently expressed in glasses, here on the earth, yea, and at the day tyme, when their lyght is eyther none or moste fe-[F3v]ble, and weake, as we see it is vsed[138] at midsommer to beholde that great starre called *Sirius,* in a glasse euen at noone days.[139]

---

[137] *Plinius* saieth] *Historia naturalis* ii. 32.

[138] used] usual.

[139] at noone daye] The Dog Star is the most brilliant of all stars.

Also we see euery night, the image of the starres in calme and quiet standing waters, then what shoulde let, but that their images myght also be expressed in watry cloudes.

Hereto may be aunswered, that the let[140] is in the cloude, which is neyther so harde as is the glasse, nor yet so continuall as the water, but consisteth of innumerable small droppes, so that except the light of the starres were stronger, it can in them expresse no vniforme images of them, as it doth in glasses, and in the water. Notwithstanding, in wryters of wonders, we reade some such like thing sometime to haue chaunsed.

There hathe been often seen manye sunnes, in the daye tyme, and after the sunne settinge, at the rysing of the full Moone, there haue appeared manye Moones, whiche was by this meanes that the same cloude, that receiued the sunne beames, in the morning, taried in the same place, and at the Moones rising, was ready also to receiue her image.

[F4r] ¶ *Of wonderfull apparitions.*

WE wil close this booke, with a brief declaration of the natural causes, of many thinges, that are seen in the ayre, very wonderfull and straunge to beholde, which in these later yeares, haue been often seen and behelde, to the great admiration[141] of all men, not without the singular prouidence of God, to forwarne vs of many daungers, that hange ouer vs, in these moste perilous tymes.

The aparition of which, as it is most wonderfull, so the serching of the cause, to vs is moste harde and difficulte. A great deale the rather, because no man hath hetherto entreprysed (to my knowledge) to seeke out any cause of them, but all men haue taken them as immediat myracles, without any naturall meane or cause to procure them.

---

[140] let] hindrance.

[141] admiration] wonder, surprise.

And I truly, do acknowledge that they ar sent of God as wonderfull signes, to declare his power, and moue vs to amendement of life, in dede miraculus, but not yet so, that they want a natural cause. For if they be wel weyghed and considered, it is not harde to finde, that they differ much from such miracles, as ar recorded in the scripture, [F4v] and admitted of diuines. So that, as I abhorre the opinion of *Epicures,* to thinke that suche thinges come by chaunce, but rather by the determined purpose of gods prouidence: so I consent[142] not with them, that suppose when any thing is deriued from any naturall cause, God the chiefe and best cause of al thynges is excluded.

Some of these wonderfull apparitions consist of circles and rainbowes, of diuerse fashions and placings, as one with in another, the edge of one touching another, on deuiding or going through another, with lyke placing of small circles, about great circles, or partes of small circles, some with the endes vpward, som downward, some asyde, and some acrosse, but all for the most part in vniforme order constituted or placed, for the order of them pleasaunt to beholde, but for the strangenes somewhat fearfull. Suche a lyke apparition, is made with the sunns or Moones images, ioyned vnto these circles, set also in good and vniforme order. The cause of these is the meting together, of all those seuerall causes, that make the circles, raynbowes, streames and images of the sunn or moone, which [F5r] ioyned altogether, make the wonderful sight of strange raynbowes, positions of circles, crosses, and diuerse lyghtes, which perteyne to the knowledge of *Optice* and *Catroptice,*[143] that teache howe by

---

[142] consent] agree.

[143] *Catroptice*] Fulke's use of these terms antedates the English examples in the *New English Dictionary*. He doubtless knew Greek and Latin treatises which used them, such as Joannes Pena's text of Euclid, *Optica & catoptrica* (Paris, 1557). The context, however, suggests acquaintance with some of the numerous discussions of optical illusions, such as those described by Giovanni Battista Porta in *Magia naturalis* (Naples, 1589).

diuerse refractions and reflections of beames, such visions are caused. So that he, whiche wyll knowe howe they are generated, must returne vnto the seuerall treatyses of raynbowes, circles, streames, and images, of the sunne or Moone, and if in them he finde not knowledge sufficient, to instructe him, I must send hym to the demonstrations of perspectiue, where he shall want nothyng.

Another sort of them, no lesse often behelde within these fewe yeares, then the former, but a great deale more straunge and wonderfull to looke vpon, are the sightes of armies fighting, in the ayre, of Castels, Cities, and Townes, with whole countries, hauing in them hills, valies, ryuers, woodes, also beastes, men, and foules, monsters, of whiche ther are no suche kyndes on the earth, and fynally all maner of things and actions, that are on the earthe, as burialles, processions, iudgementes, combates, men [F5v] women, children, horses, crownes, arms of certayne noble men, and contries, weapons of all sortes, sometymes starres, angels, as they ar painted with the image of Christ crucified, besieging of castels and townes, many thynges and gesturs done by men or beastes, the very similitude of persones knowen to the beholders, as of late, was seen the very image of the Emperour Charles,[144] insomuche that they whiche behelde it, put of their cappes, thinking verely it had been he, and of Jhon Frederick prince Elector of Saxon, who that time was prysoner with themperour. Also the image of small crosses, which hath ben not only in the ayre, but also on the earth, on mens apparell, on dishes, platters, pottes, and al other things so that the Jewes haue been full angry, that they could neither washe, nor rub them out of their aparell. In Germany, also fyers and many suche thinges, as it were long stories, seen in the ayre.

[144] Charles] Charles V, Roman emperor 1520-1558.

All these wonderfull aparitions, may be caused two maner of wayes, the one artificially, the other naturally. Artificially by certein glasses, and instruments made according to a secret part of that [F6^{r}] knowledge whiche is called *Catoptrice,* and so peraduenture some of them haue been caused, but the most part doubtlesse naturally, when the disposition of the ayre, hath been suche, that it hath receiued the image of manye thinges placed and done on the earth. And because it is apte to receyue dyuerse images, as well in one place as in another, these monstruous formes and straunge actions, or stories proceade of the ioyninge of dyuerse formes and actions, as if twoo histories, were confusely paynted in one, the whole picture would be straunge or (as the Poet sayeth,[145]) if a paynter to a mans head, should set a horses neck, and after dyuerse fethers. Sometymes also, one image is multiplied in the ayr, into many or infinite, as ar letters and crosses, whiche fill all the ayre, euen beneathe. And the light of the sunne, receiued into litle partes, maketh to apeare, as it wer many smal starres. Let this suffice, concerning these wonderfull apparitions: once agayne admonishyng the Reader, thoughe I haue enterprysed to declare these by naturall reason, yet beleuing that not somuch as on sparrow falleth to the [F6^{v}] grounde, without Gods prouidence, I doe also acknowledge Gods prouidence bryngeth these to passe, to suche ende as before I haue shewed, vsing these causes, as meanes and instrumentes to doe them.

¶ The fourth booke *of watry impressions.*

THOSE be watry impressions, that consist moste of water. In the treaty of them, are wont to be handled, these impressions, namely cloudes, rayne, dew, hore frost, hayle, snowe, springes, ryuers and the great sea it selfe.

---

[145] the Poet sayeth] *Ars poetica* 1-2.

¶ *Of cloudes.*

A cloude is a *vapor* colde and moyste, drawen out of the earth, or waters by the heate of the sunne, into the mydle region of the ayre, where by colde it is so knit together, that it hangeth vntill either the waight or some resolution[146] cause it to fall downe.

The place wherein the cloudes doe hange is sayde to be in the midle region of the ayre, because men see it is necessary that there shoulde be a colde whiche [F7r] should make those *vapors* so grosse, and thycke, whiche for the most part are drawen so thinne, from the earth, that they are inuisible, as the aier is. And although they are knowen oftentime as *Aristotle* wytnesseth,[147] to be in the lowest region of the ayre, neare to the earth, in so muche that sometymes, they fall downe to the earth, with great noyse, to the great feare of men and no lesse losse and daunger. Yet may it be reasonably thought, that these cloudes were generated in the midle region of the ayre, farre distant from the earth, whiche by their heuiness doe by litle and litle sinke downe, lower into the lowest region, and sometymes also fall downe to the earth.

The commen opinion is, that they goe not hygher then nyne myle, whiche because it leaneth to no reason[148] is vncertayne.

Albertus Magnus, whose reason also is to be doubted of, affirmeth, that the cloudes doe scarse exceade three myle in heyght, when they are hyghest.

And some let not to saye,[149] that oftentymes they ascend not past the halfe of one myle, in heyght.

---

[146] resolution] dissolution.

[147] *Aristotle* wytnesseth] *Meteorologica* 340a.

[148] it leaneth to no reason] it is not supported by reason.

[149] let not to saye] do not refrain from saying.

[F7v] Agayne other pretending to find out the truth by *Geometrical*[150] demonstrations, make it aboue fiftie myle to the place where the generation of cloudes is.

Howe these men take the distaunce from the earth, it is vncertain, whether they that assigne the least distance, meane it from the hyghest part of the earth, as are hyll toppes, or from the commen playne. Againe, whether they that assigne the hyghest distaunce to be from the lowest vallyes, of the earth, or from the hylle toppes.

The reason before shewed, moueth me to thynke that the moste vsuall and commen generation, I meane the condensation or making thick of these thinne *vapors,* into cloudes is in the midle region of the ayer. But for the distaunce of the cloudes, when thei be generated I thinke they be somtime nine mile, somtime iij. myle, somtime halfe a myle, and somtyme lesse then a quarter of a myle from the earth.

¶ *Of Mystes.*

THere be two kyndes of mistes, the one ascendinge the other descending.

That whiche ascendeth, goeth vp out of the water, or the earth as smoke, but doth [F8r] not commenly spred ouer all other parts, it is seen in ryuers and moyst places.

The other mist, that goeth down toward the earth, is when any *vapor* is lifted vp into the ayre, by the heate of the sunne, which not being strong enough to drawe it so high, that the colde maye knitte it: suffereth it after it is a lytle made thicke, to fall downe agayne, so it filleth al the ayre with the grosse *vapors,* and is called mistes, being vsually a signe of fayre weather.

¶ *Of empty cloudes.*

THere be certen cloudes, that ar empty, and send no raine, they come of ij. sortes.

[150] *Geometrical*] 1639; 1563 *Geometical.*

For one sorte are the remnantes of a cloude, that hath rayned, whiche can not be conuerted into water, for ther dryenes. Another sort is of them, that are drawen vp, of wett and drye places, and be rather *Exhalations* then *vapors,* that is they be drie, whot, and light, so that it wer harde for them to be turned into rayne. They looke whyte lyke flocks of wolle, when the lyght striketh into them. Ther be also empty clouds, when the wyndes haue dispersed abrode, any cloud, they are scatered ouer al the skie, but these cloudes, though for a time they be emptye, yet [F8ᵛ] because they consist of such a substaunce as is watrye, they may be and are often tymes, gathered together and geue plentifull rayne.

Of the collours of cloudes, we haue spoken in the second booke of fyry *Meteors,* where those collours and the causes of them, are described, whiche seeme to be fyerye, or may be thought to be inflamations or burninges, as be redde fyry and yealowysh.

But besyde those, there be whyte, black, blewe, and greane.

Whyte cloudes be thynne, and not very watery, so that the lyght receiued in them, maketh them to appeare whyte.

Black cloudes be ful of thick, grosse, and earthely matter, that maketh them looke so darke.

Blew cloudes be ful of thick, grosse, and earthly as the blacke, so the light receiued in them, maketh them to seeme blewe.

Grene cloudes are altogether watry, resolued into water, whiche receyuing into them the lyght, appear grene as water doth in a great vessell, or in the sea and ryuers.

[G1ʳ] ¶ *Of Rayne.*

AFter the generation of cloudes is wel knowen, it shall not be hard to learn, from whence the rayne commeth.

For after the matter of the cloud being drawen vp, and by cold made thick, (as is sayde before) heate followynge, which is moste commenlye of the Southerne wynde, or any other wynde of hotte temper, doth resolue it againe into water, so it falleth in droppes, to geue encrease of fruict to the earth, and moue men to geue thankes to God.

There be small showers, of small droppes, and there be great stormes of great droppes.

The showers with small drops, proceade either of the small heat, the resolueth the cloudes, or els of the great distaunce of the cloudes from the earth.

The streames with great droppes, contrariwyse, doe come of greate heate, resoluing or melting the cloude, or els of smal distaunce from the earth. Wherof we see an experiment when water is powred forth, from an highe place, the droppes are smalle, but if it be not from height, it wyll either haue no droppes [G1v] or very great.

The cause why rayne falleth in round droppes, is both, for that the partes desire the same forme that the whole hathe, whiche is round, and also that so it is best preserued against all contrary qualities, like as we see water, powred vpon drye or greasy thinges to gather it selfe into roundels, to auoyde the contrarietie of heate and dryenes.

It is not to be ommitted, that raine water although a great part of it be drawen out of the sea, yet moste commenly it is sweet and not salt.

The cause is, because it is drawne vp in suche small *vapors,* and that salt part is consumed by the heat of the sunne.

The rayn water doubtles, doth more encrease and cherishe thinges growyng on the earth, then any other water wher with they may be watered, because the rain water, reteineth much of the sunns heate in it, that is no smal comfort to all growyng plantes. The water that commeth from heauen, in rayne wyll sooner come to putrefaction, or stinking, then any other, because it hath been made very subtile by heate, and also for that it is [G2r] mixed with so many earthly and corruptible substaunces.

Rayne water that falleth in the sommer, by *Auicens* iudgement, is more holsome then other water, because it is not so colde and moist as other waters be, but whotter and lighter.

Somtime ther is salt rain, when som *Exhalation* which is whot and drie, is commixd with the *vapor* wherof the raine consisteth.

Sometime it is bitter, when some[151] burnt earthly moisture is mixed withit. This rayne is both vnholsom and also vnfruictful. In these countries, ther is great store and plenty of rayn, because the sunne is of such temperat heat, that it gathereth many *vapors,* and by immoderat heat doth not consume them. But in the East partes, in some whot contries, it neuer or seldom is seen to rain, as in Egipt and Siria, but in steade of rain Egipt hath the ryuer *Nilus,* whose ouerflowings, doth maruelously fatten the earth. In *Syria* and other like countries, they haue more plentifull dewe, then we haue, which doth likewyse make their earth excedding fruictfull.

Seneca testifieth, that the rayn soketh no deper into the earth then tenn foot depe.[152]

[G2v] ¶ *Of the signes of Rayne.*[153]

First if the skie be redd in the morning, it is a token of rayne, because these *vapors* which cause the rednes, wylbe shortly resolued into rayne.

---

[151] some] 1602; 1563 summe.

[152] tenn foot depe] *Quaestiones naturales* iii. 7.

[153] the signes of Raine] *Cf.* Sir Napier Shaw, *Manual of Meteorology* I (Cambridge, 1926): pp. 98-114; Richard Inwards, *Weather Lore* (London, 1893).

If a darke cloude be at the sunne rysing, in whiche the sunne soone after is hidde, it wyll desolue it, and rayn wyll followe. If then appeare a cloude and after *vapors* are seen to ascend vp to it, that betokeneth rayne.

If the sunne or Moone loke pale, loke for rayne.

If the sunne in the East, seem greater then commonly he appeareth, it is a signe of many *vapors* whiche will bryng rayne.

If the sunne be seen very earlye, or fewe starres appeare in the nyght, it betokeneth rayne.

The often chaunging of the wynds, also sheweth tempest.

The moste suer and certaine signe of rayne, is the southerne wynde, whiche with his warmenes, alwayes resolueth the cloudes, into rayne.

When there is no dewe at such tymes as by nature of the tyme, ther shold [G3r] be, rayne followeth, for the mater of the dewe, is turned into the matter of watrie cloudes.

If in the West about the sunne setting, there apeare a black cloude, it wyl rayne that nyghte, because that cloude shall wante heate, to disperse it.

When muche dust is raysed vp, and when the woddes make a great noyse, some tempest is towards.

Hard stoones wyl be moist and sweat against[154] rayne, lamps, and candles, by sparcling, frogges crying, trees breaking, leaues falling, and dust clottering[155] forewarne vs of a tempest.

Flees, flyes and gnats, byght sore toward a tempest, kyne feed greadely, birds seeke their vitels more besilie, for in the grosse ayer disposed to rayne, their stomack is whoter and they more hongry. But these kynde of signes perteine

---

[154] against] in expectation of.

[155] clottering] clodding, coagulating.

not so properly to *Meteorologie,*[156] as to maryners and husbandrie, which haue a great many more then these. And *Virgil* in his first booke of *Georgikes,* hath a great nomber for them, that lyst[157] to learne. Wherfore let these hetherto suffice.

[G3v] ¶ *Of monstruous or prodigious rayne.*

HEtherto we haue made mention only of naturall rayne, and that which is common whiche no man doth marueile at. But ther is somtime such rain, that worthely may be wondred at, as when it raineth, wormes, frogs, fishes, blood, milke flesh, stones, wheat, iron, wol, brick and quicksiluer. For histories make mention,[158] that at diuerse times, it hath rained such thinges, whose naturall cause for the moste parte, we will goe about to expresse, notwithstandinge accomptinge them amonge suche wonders, as God sendeth to be considered, for such endes, as we haue before declared. Wormes and frogges may thus be generated, when fat *Exhalations* ar drawen vp into the ayre by a temperature of whott and moist, such vermyn may be generated in the ayr, as they are on the earth, without copulation of male and female. Or els that with the *Exhalations* and *vapors,* their seede and egges are drawen vp, which being in the clouds brought to form, fal down among the rain.

Likewyse the spawne of fishes, being drawne vp maketh fishes to rayn out of the cloudes. The vehement heate of the [G4r] sunne, in sommer and specially in whot contries,

[156] *Meteorologie*] "In English writings the first instance of the word 'meteorology' that I have found is in William Fulke, *A Goodly Gallerye*. . . . The reference antedates the first recorded usage in the *NED* by fifty-seven years." (S. K. Heninger, *Handbook,* p. 4n.)

[157] lyst] choose.

[158] histories make mention] See the *Historia naturalis* ii. 57, 59 for examples.

draweth mylke out of the pappes of beastes and cattel, whiche being caried vp in *vapors* and resolued again into mylke, falleth downe lyke rayn.

After the same maner the sunne also from places where bloud hath ben spilt, draweth vp great quantitie of bloud, and so it rayneth bloud.

It raineth flesh, when great quantitie of bloud being drawen vp, it is clottered together and seemeth to be flesh. *Auicen* sayeth that a whole calfe fel out of the aire, and some wold make it seme credible, that of *vapors* and *Exhalations* with the power of the heauenly bodies, concurring, a calfe, myght be made in the cloudes. But I had rather thynke, that this calfe was taken vp in som storm of whirlewynd,[159] and so let fall again, then agree to so monstruous a generation.

It is a great deale more reasonable, that stones of earthly matter gathered in cloudes, shoulde be generated as we haue said befor of the thonderbolt. Yet som men thinke, that wynd in caues of the earth, breaking vpward violently, carieth before it, earth and stones into the ayre, which can not [G4^{v}] long abide, but fall downe and are compted among prodigius rayne. *Exhalations* that be earthy and drawne out of claye, haue muche grosse substaunce in them, which gathered together, and by gret heat burned in the clouds, make brick, which is no great meruayle.

He that hath seen an egges shell full of dewe, drawen vp by the sunne into the ayre, in a May morninge, wyll not thynke it incredible, that wheat and other grayne, should be drawen vp in muche whotter countries then ours is, muche rather the meale or flower whiche is lighter.

A certayne mossynes, like woll, as is vpon quinses, wyllowes, and other yonge fruictes and trees, is drawen vp of

[159] whirlewynd] 1602; 1563 whrlewynd.

the sunne, among the *vapors* and *Exhalations,* which being clottered together, falleth downe lyke lockes of wolle.

Quicksiluer all men knowe, with small heate wylbe resolued into moste thinne *vapors.* Whereof when quantitie is drawen vp, it falleth down agayne. As it is redde,[160] that once at *Rome,* it rayned quicksyluer, wherewith the brasen mony being rubbed, it looked like siluer.

[G5r] *Titus Liuius,*[161] maketh mention, that it rayned chalke, whereof the cause can not be hydde to them, that reade howe stoone and brick come into the ayre.

Iron hath also rayned out of the clouds, and sundry tymes, as histories wytnesse. Whereof this hath ben the cause. The generall matter of all metalles, which is quicksiluer, and brymstone, with[162] the speciall matter of mixtion, that maketh irone, weare all drawen vp together, and there concocted[163] into the metall, so came the straunge rayne of iron.

Auican sayeth,[164] he sawe a piece of iron that fel out of the cloudes, that weighed about an hundred pound weyght, wher of very good swerdes were afterwardes made.

¶ *Of Dewe.*

DEWE is that *vapore,* whiche in spring and *Autumne,* is drawen vp by the sunne in the daye tyme, whiche because it is not caried into the midle region of the ayre, abidyng in the lower region, by colde of the nyghte, is condensede into water, and falleth downe in verye smalle droppes.

There is commen dewe and swete dew. [G5v] One kynde of the sweet dewes, is called *Manna,* being whyghte lyke

[160] As it is redde] as is to be read.
[161] *Titus Liuius*] xxiv. 10.
[162] which . . . with] 1602; 1563 with . . . which.
[163] concocted into metal] turned into metal by the action of heat.
[164] *Auicen* sayeth] *Cf. Avicennae de congelatione et conglutinatione lapidum,* ed. E. J. Holmyard and D. G. Mandeville (Paris, 1927), pp. 24, 47.

sugar, whiche is made of thicke and clammye *vapors*, whiche maketh it so to fall thicke and whyte. It falleth onely in the East partes. As for that *Manna* which God rained to the Israelites it was[165] altogether miraculous. In *Arabia* (as *Plinius*[166] wryeth) is a very pretious kynde of dewe, that is called *Ladanum*, whiche falling vpon the herbe *Cusus*, and mixed with the iuyse of that hearbe which goates do eate, is gathered of goates heares and kept for a great treasure. Ther is another kind of swete dewes, that falleth in England called the meldewes,[167] whiche is as sweet as hony being of such substance as hony is, drawen out of sweet herbes and flowers.

There is also a bitter kynde of dewe that falleth vpon herbes, and lieth on them lyke brann or meale, namelye[168] because it is of an earthly *Exhalation*, and so remayneth, when the moyster is drawen away: this dewe kylleth herbes.

The commen dewe, dronke of cattel, doth rotte them because the matter is ful of viscositie, bringing them to a fluxion.[169]

Ther be thre thinges, that hinder dew [G6r] from falling, that is great heate, great colde, and wynde. For dewe falleth in the most temperat calme tyme.

¶ *Of hore froste.*

Hoare frost of whight frost, is nothing els but dewe congeled by ouermuche colde. The South and East wynd, doth cause dew, but the North and Northern wyndes,

---

[165] it was] 1602; 1563 was.

[166] *Plinius*] *Historia naturalis* xii. 37.

[167] meldewes] "honey-dew, a sweet sticky substance found on the leaves and stems of trees and plants, held to be excreted by aphides: formerly imagined to be in origin akin to dew" (*New English Dictionary,* Mildew *sb.* 1, quoting this passage).

[168] namelye] especially.

[169] fluxion] "a morbid discharge (of blood, excrement, etc.)" (*New English Dictionary*).

doe fryese the *vapors,* and so it becommeth hoare froste, whiche if that excessiue colde had not been, should haue turned into dewe.

The dewe and the hoare frost, agree in thre thinges, namely in matter, in qualitie of time, and place of their generation. In matter they agree, for they are both generated of a subtile and thinne *vapor,* also small in quantitie.

In qualitie of tyme they consent,[170] for both are made in a quiet and calme tyme, for if there were great wynde, it would dryue away the matter, and so cold ther be no generation.

Thyrdly they are both generated in the lowest region of the ayre, for (as *Aristotle* affirmeth[171]) vpon high hilles, ther is neither dewe nor hoare frost.

They differ also in thre things. For [G6^{v}] the hoare froste is congeled before it be turned into water, so is not the dewe.

Secondly, the dewe is generated in temperate weather, the whight froast in colde weather.

Last of all, whote wyndes, as the South and East do cause dewe, but cold wyndes as the Northe and Weast doe cause hoare froast.

Hoarefroast doth often stynke, because of the stinking matter whereof it consisteth, which is drawen out of lakes and other muddy and stinking places.

¶ *Of Hayle.*

Hayle is a hotte vapor in the mydle region of the ayre, by the cold of that region, made thicke into a cloude whiche falling downe to the soden colde of the lowest region is congeled into Ise.

There be so many kyndes of hayle, as ther be of rayn. The fashion of hayl, is sometyme round, whiche is a token

170 consent] are in harmony.
171 *Aristotle* affirmeth] *Meteorologica* 347a.

that it was generated in the mydle region of the ayre, or very neare it, for falling from hygh, the corners are worne away.

When the hayle stones are square, or thre cornered, the hayle was generated neare the earth.

[G7r] Often times, there is harde a great sounde in the cloudes, as it were of thonder, before hayle, or as it were of an army fighting etc. the cause is, that *vapors* of contrarie qualities, beinge inclosed in the cloude, doe striue to breake out, and make a noyse euen as colde water doth put into a seathing pot.

In spryng and haruest tyme, is often hayle, seldome in sommer and wynter. In wynter there wanteth whot *vapors,* in sommer, the lowest region is to whot, to congele the rayne falling downe. In spring and *Autumne,* there wanteth neither whot *vapors,* to resist the colde, nor sufficient colde to harden the droppes of that whot shower of rayne.

The haylestones are somtimes greater, and sometimes lesser: greater with greater colde, and lesser with lesser cold.

There is seldome haile in the night, for want of whot *vapors* to be drawen vp.

Sometime hayle and rayne, falleth together, when the latter end of the cloud for want of colde in the lowest region, is not congeled.

Hayle stones ar not so cleare, as Ise, because they are made of grosse and earthy [G7v] *vapors,* ise is congeled of cleare water.

Hayle is sooner resolued into water then snowe, because it is of a more soden and swyft generation.

¶ *Of Snowe.*

Snowe is a cloude congeled by greate colde, before it be perfectlye resolued from *vapors* into water.

Snowe is whyght, not of the proper colour, but by receiuing the lyghte into it, in so many small partes as in fome, or the whyghte of an egge beaten.

Snowe is often upon highe hilles, and lieth long there, because their toppes ar colde as they be neare to the mydle region of the ayre. For oftentimes it rayneth in the vallye when it snoweth on the hylles.

Snowe melting on the high hilles, and after frosen agayne, becommeth so hard that it is a stone, and is called *Christal.*[172] Other matters of snowe because they are commen wt̃h raine, are nedeles to be spoken of. To be shorte, sleet is generated euen as snowe, but of lesse colde, or els beginneth to melte in the falling.

Snowe causeth thinges growing to be fruictfull, and encrease, because the [G8r] cold dryueth heate vnto the rootes, and so cherysheth[173] the plantes.

ℂ *Of Springes and Riuers.*

THE generation of springes is in the bowels of the earth, and therfore something must be sayde of the bodye of the earth. The earth though it be solide and massy, yet hath it many hollow gutturs and vaynes, in whiche is alwayes ayer to auoyde emptines. For the ignoraunt in Philosophie must be admonished, that all thinges are full, nothing is emptye, for nature abhorreth emptines, so that where nothing els is, there is ayre and *vapors,* whiche by colde as it hath often ben sayde, wylbe resolued into droppes, as we see experience[174] in marble pillers and suche lyke harde stoones, towards raine. This aier and *vapors* therfore being turned into droppes of water, these dropps sweat[175] out of the earth, and fynde some issue at the

[172] *Christal*] *Cf.* Pliny xxxvii. 2.
[173] cherysheth] warms.
[174] experience] by experience?
[175] sweat] 1602; 1563 sweet.

length, where many beyng gathered together make great aboundaunce of water, which is called a fountayne or sprynge. The cause why suche sprynges doe runne continually, is because that aire can neuer wante in those vaines, which by colde will alwayes be [G8v] turned into water, so that as fast as the water runneth forthe, so faste is ayer agayne receyued into the place, whereby it commeth to passe, that so many springes are perpetuall, and neuer dried. But if any be dried vp, it is in a whot sommer, and such springs also they be, whose generation is not depe in the earth, and therfore the *vapors* may be made drye and the earth warme, so the spryng may fayle.

¶ There be foure kyndes of sprynges, fountains, brookes, ryuers, and lakes.

¶ *Of Fountaines.*

Fountaines be small springes, which serue for wells and conductes,[176] when there is but one place, where the water is generated, and that is not very abundaunt, ether because it is of small compasse, or small vaynes and not many.

¶ *Of Brookes.*

Brookes, boornes, or fordes, be small streames of water, that ronne in a channell, lyke a ryuer. They are caused when either the spryng occupieth a great compasse, or els two or thre small sprynges meate together in one channell.

¶ *Of Ryuers.*

Ryuers are caused by the meatynge [H1r] together, not only of many springs, but also of many broocks and fordes, which being receiued in diuerse places, as they passe

[176] conductes] conduits.

ar at the length, caried into the broad sea, for the moste part. Howbeit some riuers are swallowed vp into the earth, which perchaunce runne into the sea by some secret and vnknowen channels, some ryuers there be, that hide their heddes vnder the earth, and in another place, farre of breake out againe. They wryte also, that some ryuers being swallowed vp of the earth, in one Ilande, do runne vnder the bottome of the earth and sea, and breake forth in another Ilonde. There be also many great ryuers that run vnder the earth in great caues which neuer breake foorth. *Aristotle* sheweth of pondes and lakes, that be vnder the earth.[177] And *Seneca* speaketh of a pond that was founde by suche as digged in the earth, with fishes in it, and they that did eate of them died.[178] As eeles that be founde in darke places, as wells that haue been dammed vp etc. are poyson.

¶ *Of Lakes.*

LAkes ar made by the meting together of many ryuers, brookes and springs into one deepe valley. Whereof some are so [H1^{v}] great, that they haue the name of seas, as the great lake called *Hircane,* or *Caspian* sea. These lakes sometymes vnlade them selues into the sea, by small riuers, somtymes by passages vnder the earth.

The cause of the swiftnes of ryuers, is double, for they are swyfte either for the great aboundaunce of waters, or els because they runne downe from an hylly place, as the ryuer *Rhene* falleth downe from the toppe of wonderfull hyghe hylles.

¶ *Of whote bathes.*

SOme waters that are generated and flowe out of vaynes of brymstone, are sensybly warme, and some very

[177] vnder the earth] *Cf. Meteorologica* 351a.
[178] died] *Cf. Quaestiones naturales* iii. 19.

whott, because they runne out of whot places. These waters being also drying by nature, are wholsome for many infyrmities, specially breakyng forth of scabbs, etc. Suche are the bathes in the Weste countrye, and S. Ann. of Buckstones well in the North part of England[179] and many other els where.

¶ *Of the diuerse tastes that are perceiued in wells.*[180]

For a generall reason, the waters receyue their tast of that kynde of earth, [H2r] through whiche they runne as through a strainer. Some salt, that runn through salt vaynes of the earth, som sweet, that be well strayned, or runne through such myneralls as be of sweet taste.

Some bytter, that flowe out of such earth, as is bitter by addustion[181] or otherwyse.

Some sower or sharpe lyke vineger, which runne through vaynes of allum, coporous, or suche mineralls. *Aristotle* wryeth[182] of a well in *Scicilia,* whose water the inhabitaunts vsed for vyneger. In *Bohemia* neare to the citie called *Bilen,* is a wel that the people vse to drinke of in the mornyng, in steade of burnt wyne. And in dyuerse places of Germany, be sprynges, that tast of such sharpnes.

---

[179] England] The waters of St. Anne's Well, Buxton, Derbyshire, were known to the Romans and in the Middle Ages were very popular with pilgrims in search of health. In the late eighteenth century Buxton became a fashionable spa.

[180] wells] This and the succeeding chapter provide the most intricate problem of source in *A Goodly Gallerye.* Pliny (ii. 106) and Seneca (iii. 25) are basic. Three works only incidentally mentioned by Fulke were of great use to him. (1) Book VIII of Vitruvius's *De architectura,* the first-century classic on building, which deals with problems of water, accounting for its different qualities by the differences of the soils through which it flows. Vitruvius had been printed a dozen times by 1563. (2) Caius Julius Solinus *De situ et memorabilibus orbis,* a third-century work by a Roman imitator of Pliny, was popular in Fulke's day and of more importance to him than his single reference indicates. (3) Fulke obviously had at hand a copy of Isidore of Seville's *Origines* or *Etymologiae* (7th century).

[181] addustion] burning, scorching, parching.

[182] *Aristotle* wryteth] *Meteorologica* 359b.

Some haue the taste of wyne, as in *Paphlagonia,* is a well that maketh men dronke whiche drynke thereof:[183] whiche is because that water receyueth the fumositie of brymstone, and other mineralls through which it runneth, and so filleth the brayne as wyne doth.

¶ A recitall of such ryuers and springes, as haue maruelous effectes wherof no naturall cause can be assigned by most men, although some reason in a fewe may be founde. [H2v]

CLitumnus which maketh oxen, that drinke of it whyght, it is a ryuer or spring in Italie. *Propert.*[184] *lib. 3.* This may be the qualitie of the water,[185] very flegmatike. In *Boetia* is a ryuer called *Melas,* that maketh shepe blacke if they drinke thereof.[186]

Seneca speaketh of a ryuer that maketh redde heares.[187] These two with the fyrst, may haue some reason, that the qualitie of the water may alter complexion, and so the collor of heares may be changed, as we see in certaine diseases.

In *Libia* is a spring, that at the sunne rysing and setting, is warme, at midday colde, and at midnight very whott: this may be, by the same reason that welwater is colder in sommer then it is in wynter.[188] *Seneca* wryteth,[189] that

[183] whiche drynke thereof] *Cf. Historia naturalis* ii. 106.

[184] *Propert. lib.* 3] Although Propertius mentioned Clitumnus (ii. 19 and iii. 22), the story is more probably from Isidore XIII. xiii. 6. *Cf.* Pliny ii. 106, Vergil, *Georgics* ii. 146.

[185] water,] 1602; 1563 water ∧.

[186] drinke thereof] *Cf. Historial naturalis* ii. 106; *Quaestiones naturales* iii. 25; Vitruvius VIII. iii. 14.

[187] redde heares] Xanthus, near Ilium, but apparently not mentioned by Seneca. See *Historia naturalis* ii. 106 and Vitruvius VIII. iii. 14.

[188] wynter] the Fountain of the Sun in the country of the Troglodytae (i.e. Ethiopia) according to the *Historia naturalis* ii. 106 and Isidore XIII. xiii ("apud Garamantes").

[189] *Seneca* wryteth] *Quaestiones naturales* iii. 25.

there be ryuers, whose waters are poyson, this maye be naturally, the water running through poysonous minerals, taking much fume[190] of them. Other wells that make wodde and all thing els that be cast into them stones,[191] such welles be in Englande the cause is great colde.

Another well, maketh men madd that drinke thereof.[192] This also may haue as good reason as that whiche maketh men [H3r] dronke. As also that well which maketh men forgetful by obstruction of the brain.[193]

The same *Seneca* speaketh of a water that being dronke, prouoketh vnto lust and lechery. And why may not that qualitie be in a water, which is mixed with diuerse minealls, and kindes of earth, which is in herbes, rootes, fruict, liquors. S. *Augustin* speaketh of a well in *Egipt,* in which burning torches are quenched, and being before quenched, are lighted.[194] Among the *Garamantes* is a well so colde in the daye, that no man can abyde to drynke of it, in the nyght so whott, that none can abyde to fele it.[195]

It is incredible, that is wrytten of a well in *Scicilia,* whereof if theues did drinke they were made blynde.[196]

In *Idumea*[197] was a well, that one quarter of a yeare was troubled and moddy, the next quarter blooddy, the third green, and the fourth, cleare.[198]

Seneca wryteth[199] of another well that was six houres full and runninge ouer, and six houres decresing and

---

[190] fume] vapor.

[191] cast in them stones] *Historia naturalis* ii. 106, *Quaestiones naturales* iii. 20.

[192] that drinke thereof] *Quaestiones naturales* iii. 20.

[193] of the brain] Vitruvius VIII. iii. 22.

[194] are lighted] So reports Pliny (ii. 106). Bostock and Riley note that the story also appears in Lucretius (vi. 879 ff.).

[195] to fele it] Solinus xxix; Isidore XIII. xiii. 10.

[196] made blynde] Solinus iv, but releated of Sardinia rather than of Sicily as in Isidore XIII. xiii. 10.

[197] *Idumeo*] Edom.

[198] cleare] *Cf.* Isidore XIII. xiii. 8.

[199] *Seneca* wryteth] *Quaestiones naturales* iii. 16.

emptie, perchaunse, because it ebbed and flowed, with the sea, or some great ryuer that was neare it.

[H3v] In the hill *Anthracius,* is sayde to be a well, whiche when it is full, signifieth a fruictefull yeare, when it is scarce and emptye, a barren and deare yeare. The sufficiens of moysture, maketh fertilitie, as the wante causeth the contrary.

Men saye there is a Ryuer in Hungarye, in whiche Iron is turned into coper. Whiche may well be, seyng inke in whiche is but small coperus, and artificially myxed, of Iron, dothe conterfeyte coper in collour. In this streame maye be muche coperus, and that is naturally myxed.

Both *Seneca* and *Theophrastus,* wytnesse, that waters there be, whiche within a certayne space being dronke of sheepe, (as *Seneca*[200] sayeth[201]) of byrdes (as *Theophrastus* will haue it) changeth their collours from black to whyte, and from white to black.

Vitruuius wryteth,[202] that in *Arcadia,* is a water called *Nonacrinis* whiche no vessell of syluer, brasse, or Iron, can hold, but it breaketh in pieces, and nothynge but a mules hoofe, wyll holde it and conteyne it.

In *Illyria,* garments that are holden ouer a most cold well, ar kindled and set on fyre.[203]

[H4r]In the Ile of *Andros,* where the temple of *Bacchus* stoode, is a well that the fift day of January flowed wyne.[204]

Isidore sayeth,[205] there is a well in Italy, that healeth the woundes of the eyes.

In the Ile of *Chios,* is a well that maketh men dulwitted, that drinke therof.[206]

---

200 (as *Seneca*] 1602; 1563 as (*Seneca.*

201 *Seneca* sayeth] *Quaestiones naturales* iii. 25, citing Theophrastus.

202 *Vitruuius* wryteth] VIII. viii. 16. *Cf. Quaestiones naturales* iii. 25.

203 set on fyre] *Cf. Historia naturalis* ii. 106.

204 flowed wyne] *Ibid.*

205 *Isidore* sayeth] XIII. xiii. 2.

206 that drinke therof] *Cf.* Isidore XIII. xiii. 3.

There is another that causeth men to abhorre lust.[207] *Lechnus* a spryng of Arcadia, is good against abortions.[208]

In *Scicilia* are two sprygs of which one maketh a woman fruictfull, and the other barren.[209]

In *Sardania,* be whote welles that heale sore eyes.[210]

In an Ile of *Pontus,* the ryuer *Astares,* ouerfloweth the fieldes in whiche the sheepe that be fedde, doth geue black mylke[211]

In *Aethiopia,* is a lake, whose water is lyke oyle.[212]

Also manye sprynges of oyle haue brooken foorth of the earth, which commeth of the *viscositie* or fatnes of the same earth.

The lake *Clitorie,* in Italye, maketh men that drynke of it to abhorre wyne.[213]

[H4v] The lake *Pentasium* (as *Solinus* saith[214]) is deadly to serpentes and wholsom to men.

Seneca wryteth[215] of certeyn lakes that wyll beare men which can not swymm. And that in *Siria,* is a lake in whiche brickes do swymme, and no heuy thing wyll sinke.

It is said, that the ryuer *Rhene* in Germany wyll drowne basterd children that be cast in it, but dryue alonde[216] them that be lawfully begotten.

The ryuer *Hypanis* in *Schithia,* euery day bryngeth foorth little bladders, out of whiche flyes do come that die the same nyght.[217]

---

[207] abhorre lust] *Cf.* Isidore XIII. xiii. 4.
[208] abortions] *Cf.* Isidore XIII. xiii. 5.
[209] barren] *Ibid. Cf.* also Solinus v.
[210] sore eyes] *Cf.* Isidore XIII. xiii. 10. *Cf.* also Solinus iv.
[211] black mylke] *Cf. Historia naturalis* ii. 106.
[212] like oyle] *Cf.* Isidore XIII. xiii. 2, Solinus xxx.
[213] abhorre wyne] *Cf. Historia naturalis* xxxi. 13, Vitruvius VIII. iii. 20, Isidore XIII. xiii. 2.
[214] *Solinus* saith] v ("stagnum Petrensium").
[215] *Seneca* wryteth] *Quaestiones naturales* iii. 25.
[216] alonde] ashore.
[217] the same nyght] *Cf. Historia naturalis* xi. 43. Matrona is the Latin name of the River Marne, which is adjacent to Germany as the Romans defined it.

Matrona the ryuer of Germany, as the common people saith, neuer passeth day but he taketh some praye.

¶ *Of the Sea.*

THE sea in this treatise, hath place as a mixed substaunce, for els the element of waters being simple, were not here to be spoken of.

The sea is the naturall place of the waters, into which all ryuers and other waters, are receiued, at the length.

And here it is to be vnderstanded, that the very proper and naturall place of the [H5r] water, were to couer al the earth, for so be the elementes placed. The earth lowest, and round about the earth, the water, about the water the ayre, and about the ayre the fyre. But God the most mighty and wyse creator of all thinges, that the earth might in som partes be inhabited of men and beastes, commaunded the waters to be gathered into one place, that the drie londe might appeare, and called the drie land earth, and the gathering of waters he called seas.

In the sea are these two thinges to be considered, the saltnes, and the ebbinge and flowyng.

¶ *Of the saltnes.*

THE saltnes of the sea, accordinge to *Aristotles* mynde[218] is caused by the sunn, that draweth from it all thinne and swete *vapors,* to make rayne leauing the reste as the setling or bottom, which is salt. But men of oure tyme, peraduenture more truely, do not take this for the only and sufficient cause, to mak so great a quantitie of water salt, but say, that the sea by Gods wysdome is gathered into such valleys of the earth, as were other wyse barren and vnfruictful, such earthes [H5v] are salte, the sea water then

[218] *Aristotles* mynde] *Meteorologica* 358a.

mixed with that earth, must needes be salt, els ryuers by *Aristotles* mynde, should be salt as well as the sea. The Reader maye chuse whiche opinion is most probable.

¶ *Of the ebbing and flowing.*

THE ebbing and flowing of the sea, as *Aristotle* semeth to teach, is by reason of *Exhalations,* that be vnder the water, whiche dryue it to and fro, according to contrary boundes, and limites, as vpwarde and downwarde, or[219] wyde and narrowe, deepe and shallowe. This opinion of *Aristotle* also, as more subtile then true experience teacheth men to mislike and to ascribe the cause of ebbing and flowyng to the course of the moone, which ruleth ouer moysture, as the sunne doth ouer heate, for from the new Moone to the full, all humors do encrease, and from the ful to the newe moone, decrease agayne. Also the very true tyme of the ebbing and flowyng may be knowen, by the course of the Moone, with whome as the ladye of moysture, we will close vp the fourth booke of moiste and watry impressions.

[H6r] ❧ The fift booke of earthly *Meteores* or bodies perfectly mixed.[220]

THIS last treatise conteyneth suche bodies whose chief matter is the earth, and are called perfectly mixed be-

---

[219] or] ed.; 1563 of; 1602 om.

[220] bodies perfectly mixed] For his account of earths, liquors concrete, metals, stones Fulke unquestionably read Cardan's *De subtilitate,* books V-VII, and perhaps *De rerum varietate* (see introduction), although the discussions in the latter are relatively brief and general. Very possibly he knew Albertus Magnus, *De rebus metallicis et mineralibus,* in which the conception of "digestion" is prominent; this idea, however, he may have derived from the alchemical literature, with which he evidently had some acquaintance. Although he mentions Pliny, he does not owe a great deal to him in this section. He does not appear to have known Agricola *De re metallica* (1556), as is rather surprising, nor could he have known the same author's *De natura fossilium* (1566), which makes use of almost the same organization as Fulke does. See *De re metallica,* tr. Herbert and Lou Henry Hoover (London, 1912), pp. 1-3. The notes and appendices to the Hoovers' edition of Agricola give an admirably full account of the writings on mineralogy prior to 1556, including the chief alchemical works.

cause they ar not easly resolued in to the chiefe matter, wherof they ar generated. These are deuided into foure kyndes. The first be diuerse sortes of earth, the second be liquors concreat,[221] the third be metalls and metalliques, the fourthe be stoones. This deuision is not altogther perfect both for that ther be many of these mineralls whiche partake of two kyndes, and also for that the names of some of these kyndes may be sayde of other. Yet mindyng as plainly as can be, to declare the thynges them selues, the controuersye and cauillation of[222] names, shall not greatly trouble vs. Especially seyng we pretende not to teache Philosophers, but such as nede a ruder and plainer instruction. They may therfor be content with this diuision, which shal not serue them to dispute of these matters, but to vnderstand the truth of these [H6^{v}] thinges that they desyre. Of these fowr, therfore we will speake orderly and generally, not mynding to intreate of euery particuler kynde (for that were infinit) but to open suche vniuersall causes, as they whiche haue witte, may learne, (if they list) to apply vnto al particulers.

¶ *Of earthes.*

THe earth is an element, one of the foure, cold and drie, moste grosse and solyde, moste heauy and weighty, the lowest of all other in place. When I saye an element, I meane a simple body vncompounded. This earth is no *Meteore,* but as it was shewed in the water, to the end ther should be generation of things, there is no element that we can haue, whiche is pure and symple, but all are mixed and compounde. Our fyre is grosse and compounde, so is our ayer our water, and our earth. But the earth notably and aboue the rest is mixed. For the puer and naturall earth is drye and cold, but we see much to be moist, and much to be hoat. The naturall earth is blacke of collor, but we see

[221] concreat] solidified.
[222] cauillation of] quibbling over.

many earthes white many yelowe, and many redde. So that first the greatest part of the earth is mixed [H7r] with water, that maketh it to cleaue to gether, with ayer and some fyre, which make an oyly fatt or claymy earth, as is claye made, etc. Another great part is dryed not into the natural drienes of the first qualitie, but as a thing ones mixed and after dryed, ether by the cold, as sand grauell, etc. or els by heate, as chalke, oker etc. And yet somwhat more plainly and particularly to discourse vpon these causes, admitting the naturall collour of the earth to be black, of the water to be blewe, of the ayre to be whyte, and of the fyre to be ruddy, it followeth that vpon the mixtion of these collours, or chief domination of them, al thinges hath their collour. The grosse substaunce of the earth therefore beinge diuersly myxed with other elementes, and those myxtures againe being eftsones[223] altered, by dyuerse and sometime contrary qualities, hath brought forth so manye kyndes of earth, as claye, marle, chalke, sand, grauell etc. Claye is mixed with fat moisture takyng his colloure of the mixture with redde from whyte, but beyng colde, it is not so fructfull as marle, whiche is not alwayes so moiste as it. Chalke is an [H7v] earth by heat concocted, after diuerse mixtions and dried vp. Oker both yelow and redde with suche like are of the same nature with mixtion of redde more or lesse.

Sande and grauell are dried erthes, as it were froasen by colde, grauell is grosse and apparent, sand though it be finer, is of the same generation consisting of many small bodies, which are congeled into stones. Sand semeth to be clay dried by cold and coacted together into small stones, wherof some ar through shining which were the moyst partes, the thicke were of the grosse parte. The same is grauel, but of greater stoones consisting. The lyke iudgement is to be geuen of all other kyndes of earth, whose generation by the similitude

[223] eftsones] soon.

of these, wyll not be very harde to fynde out. They that lyst to knowe the diuerse kyndes of earthes, must haue recourse to *Plinius, Cardane,*[224] and other wryters, that recite a great nomber of them, but these are the chief and most commen kyndes.

¶ *Of liquors concreat.*

VVE take not lyquors concreat so largely, as the worde dothe signifie, for than should we comprehende, bothe [H8^r] the other kyndes followyng. But onely those liquors, called in latin *Succi,* which are as it were midle betwene metals and stones, of whiche some being fat and oyly, do burne, as brimstone, seecoles, geate, *bitumen,* etc. and the kyndes of all these. other some doth not burne, as salt, alum, coperus, saltpeter, etc. and the kindes of these. Of the first sort, which are generated of earthy and ayry *vapors,* fumes and *Exhalations,* the chief and most notable, is brimstone, which semeth to be the matter of all drie and whot qualities, that ar in earthly *Meteores.* The rest are generated of such lyke *vapors* as brymstone is, but then they be diuersly mixed. As the coles, haue much earth mixed with brimstone. Gette, seemeth to be all one, but better concocted then coles. Of amber is great contention whether it be a mineral, or the sperme of an whale, for it is found in the sea, cast vp on the shore. Now the whales seede, being of the very same qualities, is taken more and lesse concreate of diuers hardnes, som almost as hard as amber, som softer, and som liquid. Yet *Cardan*[225] plainly defineth, that amber is a minerall. Whether he haue reason or experience, contrary to the vulgar [H8^v] opinion, let them consider that list

---

[224] *Cardane*] *Historia naturalis* xvii. 4; *De subtilitate* ii (*Opera omnia* **3:** pp 404 ff.).

[225] *Cardan*] *De subtilitate* v (*Opera omnia* **3:** p. 444) and vi (iii. 457). See also *De varietate rerum* xvi (**3:** p. 307), which is more explicit and the chief reason to believe that Fulke was acquainted with this work.

to contende. These minerals that will resolue with fyre, it is apparent, that they were concreat with colde. In that they burne it is manifest, they haue a fatte and clammy substaunce, mixed with them. As the other kynde hath not, whiche wyll not resolue so well with fyre, as with water, whiche be salt, coperus, saltpeters, etc. these burne not being watry, earthy, and not fatt, vnctuus, nor clammy.

These be of diuerse collours, black as coles and geat, because ther is much earthy substaunce mixed with their sulphureus matter. Some be shere[226] as saltt and alume, hauing a substaunce watry, dryed, and concreat. Coperus is greane, because it hath muche colde matter that is blewe, mixed with it. Salt the most commen and necessary of all these liquors concreat, that be moist and not fatty, hath two maner of generations, one naturall, and the other artificiall. The naturall generation, is when it is first generated, in the earth, after commeth the water of the sea, and is infected with it, out of whiche the salt is againe artificially gathered. Of these liquors concreat [I1r] be those strange wells and sprynges infected, of whiche was spoken, in the latter ende of the fourth booke. Most notably brimstone causeth the whot bathes, and burneth in *aetna,* of *Scicilia,* and *Vesuuius* of Italye, casting vp the pumise stones, of whiche is no place here to entreate.

¶ *Of Metalles.*

METalles be substaunces perfectlye myxed, that wyll melte, with heate, and be brought into all manner of fashions that a man wyl. Of these the Alcumistes saye, there be seuen kyndes, to aunswere to the seuen Planetes. Gold, syluer, copper, tynne, lead, Iron and quicksyluer, that they cal *Mercury.* But sauing[227] their authorities, quick-

---

[226] shere] bright, shiny.
[227] sauing] with all respect to.

syluer is no more a metall, then brymstone, whiche is as necessarye to the generation of metall, as quicsyluer is. For they all agree, that all metalles are generated of sulphur, that is brymstone, whiche because it is whot, they call the father, and *Mercury* that is quicksiluer, which because it is moyst, they call the mother. So by as good reason, may they call brymstone a metalle, as *Mercury*. Then there remai-[I1ᵛ] neth but six perfect metalls, Gold, Syluer, Copper, Tinne, Lead, and Iron.

¶ *Of Gold.*

THat moste vnprofitable and hurtfull of al metalls golde, which most men disprayse, and yet all men would haue, is of all other metalles the rarest, it is only perfect, all other be corruptible. Gold neuer corrupteth by rust, because it is pure from poysonus infection and most solide, that it receiueth not the ayre into it, which causeth all thinges to corrupt. It is perfectly concocted with sufficient heate, and mixture of Sulphur, all other metalls, either are not so well concocted, or els they haue not the due quantitie of brimstone. This opinion hath also place among the Alcumistes, that because nature in al her workes, seketh the best ende, she entendeth, of al metals to make gold, but being let[228] either for wante of good mixture, or good concoction, she bringeth forth other metals, in deede not so precius, but much more profitable, and the lesse pretius, the more profitable, for ther is more vse to the necessitie of mannes life, in Iron and lead, then is in golde, and syluer. But either the bewtie, or the per-[I2ʳ] fection, or at lestwyse the rarenes of gold and siluer, haue obteined the estimation of al men, so that for them is sold al maner of things, holy, and prophane, bodely and spirituall. What paynes doth not men take to wynn gold? euery man hath one way or other, to hunt

[228] let] hindered.

after it, but the Alcumist despising all other wayes as slow, vnnaturall, and vnprofitable, laboreth ether to helpe nature in her worke, as of vnperfect metals to make perfect, or els to force natur to his purpose, by his *quintessences* and *elixers,* so that what by purging, what by concocting, what by mixing of sulphur and quicksiluer, and muche other like stuffe, at length he turneth the wrong side of his gowne outward, all the teeth out of his head, and his body from helth to a palsey, and then he is a Philosopher, and so he will be called.

¶ *Of Syluer.*

Siluer the most pure metalle, next vnto golde, hath indifferent good concoction in the earth, but it wanteth sufficient heat in the mixtur, that maketh it pale. It is founde as they saye, running into diuerse vaines as all other metalles be, but this most specially, after the shape and fashion [I2v] of a tree, lying alonge with a bodye or stocke of proportion lyke to the body of a tree, also with armes, braunches, leaues and fruictes. This metall syluer, lacketh sufficient heate, and therfore commeth neyther to the collour, soliditie, nor perfection of golde, and is generated in colde countries, neare vnto the North, and South poles. In so greate quantitie, that hasbandmen, when they plowe the grounde turne vp syluer, among the clottes[229] in their dayly labours. Whiche they doe hyde, and conceale, least the gready Princes, for couetousnes of the metall, should ouerturne and destroye their lande. The golde mynes, are contrarywyse, moste founde in the whote countries of *India* and *Aethiopia,* because in them is sufficient of heate for that vnhappy generation.

---

[229] clottes] clods.

This syluer also, the Alcumistes woulde fayne make by arte, but *Mercury* the chief maister of the worke, is so subtill, and so slye, that nothinge can holde hym, nothing can kyll hym. For if the glasse be not very thyck, he wyll soone breake out of pryson, and so there is nothyng left.

[I3r] ¶ *Of Copper.*

Copper in collour, comming nearest to golde, beyng not solyde, nor massy, (for of all metall golde is the heauiest) geueth waye to corruption, beyng infected with that greane minerall copperus. Hereof be dyuerse kyndes, brasse, latine and suche lyke, which differ in digestion, the copper beyng purest, is of best digestion and nearest vnto golde, and so the rest in lyke degrees. Copper is moste lyke to syluer in the wayghte, and in the hammeryng, wherefore the Alcumistes haue learned to make it whyte, that it deceyueth mens syghte and handlyng, but the Goldsmythes doe easely trye it and by the teast of counterfect siluer, maketh copper agayne. Copper or brasse, doth alwaye growe neare vnto the myne of copperus, whiche runnyng with it in the digestion or naturall concoction, hyndreth it of perfection, maketh it to stynke, and to be eaten of a greane rust. Muche a doe the Alcumistes haue to turne it into golde, if it might be, they dispute very reasonablye, and conclude almoste necessarily in their talke, that it may be conuerted into golde as, a body, [I3v] that wanteth litle of perfection which may be easely added vnto it. But in conclusion of the worke, it is an harder matter, to bryng it to passe, then it was to purpose[230] before they had done it, to builde an abbay at euery myles ende, vpon Salisbury playne, as one was mynded.

[230] purpose] propose.

¶ *Of Tinne.*

Tynne, wherof great plenty groweth in the West partes of Englande, in bewty and collour commeth nearest to siluer, and of siluer wanteth nothing, but soliditie and hardnes. For tinn is a rawe and vndigested metall, also very porose and vncompact, which causeth it to crashe,[231] when it is broken or bitten. So it faileth of heat, in the commixtion and also sufficient digestion in the earth. Otherwise it is a fayre and profitable metal, to serue the vse of them, vnto whom siluer and gold are not so plentiful.

¶ *Of Lead.*

Lead also found in great abundance within this realm, is a rawe and vndigested metal, as tin is, but yet of better digestion then commixtion. For it is mixed, with a grosse earthy substance, which maketh it to be in collor so black, and so fowl to corrupt. So that of the same fumes and *exhalations,* which [I4r] if they had ben pure and well digested, if the place and matter wold haue suffered, shold haue ben concreat into siluer, for lack of the same, lead is generated, which comming plentifully, doth better seruice then syluer.

¶ *Of Iron.*

Iron the most necessary and profitable of all other metalls, and yet as ill vsed of many as any other, is generated of such substaunce as syluer is, but myxed with a redde minerall, whiche eateth it with redde ruste, and also being of two extreme degestion, passing[232] all other metalls in hardenes. And as other metals to the perfection of syluer, wante sufficient concoction, wherby they comme not to the same hardenes: so Iron passeth and exceadeth syluer is im-

---

[231] crashe] make a crackling noise.
[232] passing] surpassing.

moderate digestion. But though it come not to the perfection of syluer, God forbidde that al Iron had beene tourned into syluer, for then we should more haue myssed it, then syluer or golde, the want of whiche would hinder vs nothyng at al.

ℭ *Of Quicksiluer.*

THough quicksyluer, be no metal yet because it is the mother of al metals, some thynge is here to be spoken of it.

[I4v] There be diuerse and sondrie opinions, both of the generation, and also the qualities of it, whiche make the generation to be harde to fynde out. For if the qualitie were certainly agreed vpon, there were an easier waye founde, to trye out the generation. Some affirme, that it is exceading whott, and that they wolde proue, by the swyfte percing ther of into other thynges, that be porose. Other saye, it is exceadyng colde, and that they proue, by the exceadyng weyght of it. As for the percyng, they saye: it is[233] caused of the exceading moystnes, of which qualitie both partes doe graunte that it is. Concerning the generation, some haue sayde that it is pure and elementall water, some agayne hath thought, that it droppeth out of heauen, and is a part of the heauenly substaunce. And other sayde, that it is generated in the cloudes, and falleth downe in the field, in a circle, on those round circles, which are seen in many fieldes, that ignoraunt people affirme to be the rynges of the fayries danses. It is certayne, that quick syluer hath dyuerse tymes fallen out of the cloudes, as we haue declared in the [I5r] treatyse of wonderfull and merueylous rayne, but whether it so fall in circles, it is doubtfull. The moste probable opinion is, that it is generated of moyste *vapors* of the earth, coacted by cold, much lyke to water, as brymstone is of

[233] it is] 1602; 1563 is.

hotte fumes coacted[234] by colde, muche lyke to fyer. And thus muche of metalles.

¶ *Of Stones.*

STones the fourth kynde of earthlye myxed bodies, haue two maner of generations, by moste contrarie qualities. For heate doth harden moyst bodyes in to stones, as we see that of claye it maketh exceadyng harde brycke.

Also the thonderboltes in the cloudes, are generated by heate, as before hathe ben shewed. But colde dothe by congelyng, generate, many more stones then heate doth: for the moste parte of all the stones that are digged out of the earth, are generated by colde, whiche is able to conuerte any other kynde of myxed substaunce into stone, as hath been partly shewed in the nature of welles, and sprynges, of whiche there be some in Englande, which by their colde turne wodde or any lyke thyng into stones. I [I5ᵛ] haue seen a peece of rotten wodde, which to sight was very light, and like wodde, but in handling a very stoone that was taken out of suche a well. Also of other thynges taken out of the earth turned into stones, I haue seen and founde my selfe, flyes, with heade and wynges, very harde stones, also I haue seen an hart, a byrds tongue, a beastes stone, a peare, a plomme, and dyuerse other things turned into hard stones.

Of the diuerse kindes of stones.

STones may first be deuided into rude and bewtifull, the rude conteine those great rockes, whiche are generated by many smal partes ioyned together, and the commen peble stones, that be found euery where in the earthe, among grauell and on the shore of the sea, or bankes of the ryuers.

---

234 coacted] concentrated, compacted.

These are generated of grosse and earthly humors, congeled by colde, and because they be neyther fayre of collour, nor through shyning and also commen, they are contemptible. The fayre or bewtiful stones be either great or small. The great be, as marble of diuerse kyndes, and collors, alabaster and suche lyke which being hard and wel concocted, [I6r] may be poolished and becom bewtiful.[235] Their collor is as they are mixed, being vncongeled, so is their purenes. The small are more pretius, and they be either thick or pellucide. The thicke, be nether so fayre nor so pretius, as the *Achates,* the *Iasper, Prassios*[236] etc. These consisting of a pure matter and not very watry, are congeled into such stones. The clear stones be liquors concreet, as the *Diamond,* the *Saphir,* the *Emeralde,* etc. they ar praysed of their gretnes, hardnes, clearnes, and faire collors, of which, enough hath been spoken. Sauing that som be of opinion, that these be generated by heat, because the best are found in whot countries, in the east, and in the south. Answer may be made, that the whotter the ayre is the colder is the earth, so that reason is of smal force.

ℭ *Of the vertue of stones.*

Some perchance, would loke that we shold make a long discourse of the vertue of stones, and wold be well content that we shold entreate of diuerse properties of *gemmes* and pretious stones. Which matter though it be out of our purpose (whiche considereth only the generation) yet seing it is not out of their expectation som thing briefly and yet [I6v] sufficiently shalbe saide, of the vertue of stones.

That vertue that is ascribed vnto them, is eyther naturall or magicall. Natural vertue is either that whiche is knowen

---

[235] bewtiful] 1602 beautifull; 1563 bewitful.

[236] *Prassios*] prase, a kind of green quartz; *cf.* Albertus Magnus, *Liber mineralium* ii. 1, 8, 14.

to haue a naturall cause, or a naturall effect, as the *Magnes,* or loadstone to drawe iron, whiche is by a similitude of nature, and suche an appetite, as is betwene the male and the female. Also the sayde *Magnes,* moueth towarde the North, and as some saye, there is an other kynde founde in the Southe, that draweth towarde the south. They saye, that there is great hylles of this stone, in the North and South, which maketh it looke that waye.

Other bryng a *Mathematicall* reason, whiche because it is more curius then can be vnderstoode of the commen sort, not exercised in *Geometrie,* I omitte.

The gette and amber draw heares, chaffe, and lyke light matter, but beyng before chaffed, for heate is *attractiue.* Also the precious stone called *Astroites,* moueth of it self in vineger, the sharpenes of the vineger, percing it, and the ayer excluded, driuing it forward.[237] These [17r] vertues because I haue seen, I haue set for an example, generally all other lyke naturall vertues, proceade of lyke naturall causes, which by their effect the ingenious must seeke to fynde out. As for *Magicall* vertues be they, whiche are grounded of no reason, or natural cause whiche if they take effect, it is rather of the superstition and credulitie of hym that vseth them, then of the vertue of the stones. As that an Emerald encreaseth loue, a Saphir fauoure, a Diamonde strength, and such lyke vertues of which *Albertus* in his age surnamed the great, tooke paynes to wryte a booke, whiche I suppose be englished.[238] To conclude with the cause why stones melte

---

[237] forward] *De subtilitate* vii (*Opera omnia* **3:** p. 465).

[238] englished] *The boke of secretes of Albertus Magnus, of the vertues of Herbes, stones and certaine beastes.* This book exists in three editions printed by William Copland (active between 1548 and 1568 or 1569), none of which can be dated accurately. I have used *S.T.C.* 260. Of the sapphire it says (sig. Dviiv f.): "¶ If thou wylt make peace. * Take the ston, which is called a Saphire, which cometh from the Easte into Inde, and it that is of yelowe coloure is best, whyche is not verye bryghte, it maketh peace and concorde, it maketh the mynde pure and deuoute toward God, it strengtheneth the mynde in good thynges, and maketh a man to cole from inward heate."

not as metalles doe, may be gathered by that which hath been sayde before, because they are congeled past that degree, and also because there is left in them no vnctuus, or clammy matter. Let this suffise for stones, and so the whole purpose is at an ende.

FINIS. *W.F.*

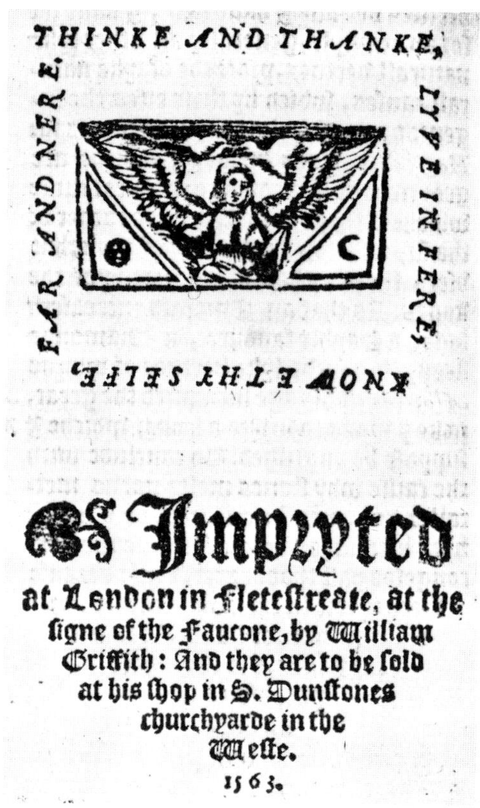

THINKE AND THANKE,
LYVE IN FERE,
KNOW THY SELFE,
FAR AND NERE.

Imprynted
at London in Fletestreate, at the
signe of the Faucone, by William
Griffith: And they are to be sold
at his shop in S. Dunstones
churchyarde in the
Weste.
1563.

With permission of The Huntingdon Library.